The Third-Quarter Century
of the
American Accounting Association
Association
1966-1991

Contents

Chapter 6: Educational Involvement

Chapter 7: Educational Research and Reforms

Chapter 8: Accounting Research

Chapter 9: Summary 181

FOREWORD

For the celebration of the 50th anniversary of the American Accounting Association, the 1965 Executive Committee decided to publish a suitable commemorative document. The result was the publication of *The American Accounting Association, Its First 50 Years*, authored by Stephen A. Zeff.

The 1989-90 Executive Committee decided to publish an equally suitable document as a part of the celebration of the Association's 75th anniversary. Thomas J. Burns, Chairperson, Stephen A. Zeff, and Thomas J. Nessinger were charged with the responsibility of selecting an author for the project and to supervise the execution of the project to assure that a monograph was published in time for the August 1991 AAA annual meeting. The committee wisely selected Dale L. Flesher to do the research and write the monograph. Thanks primarily to his efforts, with the assistance and support of the committee, he completed the project in a high-quality manner before the deadline.

It is not surprising that more pages were needed to summarize the third quarter century of the Association's activities than for the entire first half century. The past twenty-five years have been a period of rapidly-expanding activities as evidenced by the formation of special sections, the growing importance of regions, and the beginning of conferences, consortia, and seminars. The Association's publication efforts have expanded dramatically from the publication of *The Accounting Review* in 1966 to the publication of five section and three Association-wide journals in 1991. There was no newsletter in 1966. Now there are a dozen. The expansion of other activities is equally impressive. Much of this expansion has been made possible by hiring our first Executive Director, Paul L. Gerhardt, twenty-five years ago, and establishing a permanent administrative staff in Sarasota, Florida. Dale has effectively captured these changes and many more in the nine chapters of *The Third-Quarter Century of the American Accounting Association; 1966-1991*

The Association is indebted to the author and his committee, and to Pat Calomeris of the Sarasota office for her editorial and production assistance. These efforts and the resultant publication are an important part of the celebration of the AAA's 75th anniversary.

Alvin A. Arens
President, 1990-1991

PREFACE

The completion of this work coincides with the 75th anniversary of the American Accounting Association—the premiere organization in the world for accounting educators. This volume is intended as an extension of the 50th anniversary history published in 1966, and thus does not overlap any material found in the earlier history.

The author wishes to thank the many people who have contributed to the success of this project. The initial idea came from former director of education Gary J. Previts and 1988-89 president Gerhard Mueller. Subsequent presidents John Simmons and Alvin Arens have continued to support the project. In addition, other presidents who served during the past 25 years have contributed to the project through lengthy interviews. Every living past president (only Lawrence Vance is deceased) was interviewed, as were many other national, regional and section officers. In addition to personal interviews, hundreds of letters were sent to past regional and section officers asking for information about activities of long ago. All were most cooperative in their responses.

Terry K. Sheldahl of Savannah State College contributed to the success of this project in a number of ways. Sheldahl conducted about half of the interviews with the presidents and was largely responsible for the development of the twelve-page questionnaire that was used as an interview tool. He will soon complete a detailed study of AAA central governance and administration, 1966-91, and plans a second book covering research and domestic outreach activities of the era, both of which will be placed with an outside publisher.

Much credit must go to Paul Gerhardt and the rest of the loyal and devoted staff at AAA headquarters in Sarasota, Florida. In particular, Pat Calomeris, Marie Hamilton, Karen Nuss, Linda Sydenstricker, Debbie Gardner, Pam Paulson, and Sara Laurie were most friendly and helpful to a researcher who was constantly in their way and asking them questions. The staff at headquarters is quite productive, and Gerhardt deserves credit as an administrator for hiring such highly qualified people; for an organization with over 10,000 members, that publishes eight journals and twelve newsletters, the AAA has only 12 paid employees. These people deserve much acclaim for all that they have been able to accomplish on behalf of the membership.

The 75th Anniversary History Committee, chaired by Thomas J. Burns, deserves commendation for its contributions to this project. Besides Burns, Stephen A. Zeff and Thomas Nessinger served on the committee. Since this volume is essentially an extension of Zeff's 50th Anniversary History, it was most appropriate that he serve on the committee.

My employer, the School of Accountancy at the University of Mississippi, contributed secretarial and other support throughout the two years of the project.

Lastly, I want to thank my wife, Tonya K. Flesher, and my son Flyn for giving me the time and support to complete this project. As a fellow accounting academician, Tonya provided many insights which have found their way into this volume.

Dale L. Flesher, April, 1991

CHAPTER 1
AN INTRODUCTION TO AAA'S THIRD QUARTER CENTURY

The American Accounting Association (AAA) celebrated its 50th anniversary in 1966 at the Deauville Hotel in Miami, Florida, with Herbert E. Miller as the 50th president of the organization. In many ways, that 1966 meeting marked the culmination of one era and the beginning of a new era for the organization. In that year, the AAA hired a full-time executive secretary and opened a permanent office in Evanston, Illinois. Recently, Alvin Arens, the 1990-91 AAA president, called the hiring of Paul Gerhardt as executive secretary the most important event in the organization's history [Arens, 1990, p. 92]. Arens was probably correct. Paul Gerhardt has remained with AAA throughout the past quarter century, and has served the organization effectively. Without exception, all of the past presidents have empha-

sized the major role that Gerhardt has played. In a sense, this volume serves as a monument to Gerhardt since it covers 25 years, the same period that he has been with AAA. He has been involved in every aspect of the organization throughout that time, and has seen the organizational role grow many fold.

Paul L. Gerhardt graduated from the University of Colorado. In 1966, he was managing a small professional association in Chicago. The Executive Committee had advertised the position in the Chicago newspaper because that city was centrally located, and close to a large percentage of the Association's membership. Hired at a salary of $10,500, Paul Gerhardt took of-

Paul L. Gerhardt

fice on April 1, 1966. Gerhardt's first major task was to lease an office in the Chicago area. Evanston was selected as the site for locating AAA's first permanent office. Gerhardt thinks that Evanston appealed to the Executive Committee because the city was the home of Northwestern University. On June 15, 1966, the Evanston office was opened. Gerhardt hired three secretaries (one part-time) and a part-time stockboy; purchased desks and chairs from the University of Iowa at a 60-percent discount, and obtained other furniture [Minutes, August, 1966, p. 2]

Gerhardt was also responsible for the Association's move to Sarasota in 1971. Gerhardt reported at the December 1970 Executive Committee meeting that the

Evanston facility was becoming crowded and expensive, and suggested that a move to a new location might be in order in the near future. At about the same time, Gerhardt received a job offer from another association that would have meant a 50% increase in pay and the opportunity to move to Champaign-Urbana. Since he was growing weary of the Chicago winters and the stresses of raising a family in the big city, Gerhardt was tempted by the offer. However, the Executive Committee did not want to lose Gerhardt and a counter offer was made. Although AAA could not match the salary offer, it did give Gerhardt about a 25% salary increase and the option to move the headquarters to whatever location he desired [Gerhardt, 1991]. Gerhardt had visited Tampa and other cities along the Florida Gulf Coast for the purposes of determining their suitability as a site for a future AAA annual meeting. Gerhardt decided that he liked the Gulf Coast of Florida, and soon found a building that the Association could rent at 653 South Orange Avenue in Sarasota. The rent and other office costs would be much less than would be the case in Evanston. However, the building owner hesitated to rent to the AAA because he first wanted to try to sell the building. Gerhardt asked members of the Executive Committee if AAA would want to purchase the building. The response was negative. Then, after first checking with selected members of the Executive Committee for approval, Gerhardt offered to personally buy the building and rent it to the Association at a rental rate lower than the going market rate. Thus, AAA and Paul Gerhardt moved to Sarasota in the fall of 1971. Gerhardt recalls that the 1971-72 Executive Committee had a change of heart about the desirability of renting from its executive secretary, so he sold the building (at a loss) in 1972. One of Gerhardt's first acts in Florida was to hire Marie Hamilton as office manager. Marie is now celebrating 20 years with AAA.

Following a period of time at 653 South Orange Avenue, the Association eventually built its own building in 1978 at 5717 Bessie Drive. For those who have never been to Sarasota, Bessie Drive is not the easiest street to find. In fact, the AAA at 5717 Bessie Drive is the only address on that street. Given that Bessie Drive is a very short street, most people in Sarasota have never heard of it. Still, the location is quite convenient and offers low costs and the availability of a high quality labor supply. The AAA's building is fully paid for and even today offers adequate space for present and short-term future needs.

The idea of buying or constructing a new building arose in 1977, at the suggestion of past president Wilton Anderson. At the August 1977 meeting, the Executive Committee voted to ac-

Marie Hamilton

quire land near the corner of Beneva Road and Clark Road on the outskirts of Sarasota, for approximately a dollar per square foot. Gerhardt was asked to hire an architect to draw up plans for a new building. The land and building were financed primarily through mutual funds carried as investments. The one-story building opened December 1, 1978, and provided much more space than the previous headquarters. In 1981 the Executive Committee decided to enlarge the building by 2,280 square feet, at a cost of $105,000 [Minutes, March, 1981, p. 31]. Construction was completed in January 1982. The setting on Bessie Drive is absolutely idyllic in that the building is located on the banks of a beautiful lake. Employees can look out their office windows and see ducks swimming by and, occasionally, fish jumping.

AAA Leadership

The volunteer leadership of AAA over the past quarter century is summarized in Exhibit 1-1, which lists all members of the Executive Committees during the period. The Executive Committee is the primary governing body of the organization. Although the presidents have been important in the development of the organization, they have not been as important as the Executive Committees of which they were a part. All of the past presidents were asked whether an inefficient or otherwise bad president could harm the organization. The consensus was that any risk of harm was minimal because the Executive Committee is a sufficient check on the president. At the same time, this check may limit the contributions that could be made by an otherwise outstanding president. In a sense, people become president because of their past contributions to the AAA, and as president they have to be willing to allow others to make contributions. Throughout this volume, the contributions of various past presidents will be noted, but in most cases those contributions have been made over a number of years—not just during the year as president.

For those concerned about the representation of minorities and women, there have been two blacks and four women who have served on the various Executive Committees. William Campfield and Sybil Mobley were the two blacks; the four women were Catherine Miles, Mobley, Corine Norgaard, and Wanda Wallace. The latter two are currently serving. This is not a particularly high ratio of representation for either group, but not unduly low given the small numbers of female and black professors until recently. Two Canadians have served—Ross Skinner and J. E. Smythe.

As can be observed from the list, there have been minor changes in the organizational structure over the years. The current executive committee consists of the following ten positions:

President	Secretary-Treasurer
President-Elect	Director of Research
Past President	Director of Education
Vice Presidents (3)	Director of Publications

This is slightly different than the composition of the 1966-67 eleven-member Executive Committee which did not include the director of education nor the director of publications. The latter position was represented by the editor of *The Accounting Review*. In addition, there were two past presidents in 1966 and four vice presidents. The number of vice presidential positions was reduced to two in 1970. Since the director of education had been added in 1969, the 1970-71 Executive Committee had been reduced to ten members. In 1981, a third vice president was added when a by-laws change eliminated the second past president from the Executive Committee [Minutes, August, 1980]. The latter position was eliminated in order to reduce the length of time that a leader had to commit to the organization, and to infuse new blood into leadership positions.

Members of Executive Committee: 1966-1991

In 1986, the position of director of publications was added to the Executive Committee to replace the editor of *The Accounting Review*. Since the Association had instituted two new journals in 1986, it was thought inappropriate to have only one of the three journal editors on the Executive Committee. Thus, the directorship of publications was established to oversee the activities of all of the AAA journals. The first director of publications (1986-87) was Robert Sprouse, which marked the first time in many years that a past president had returned to the Executive Committee. Interestingly, Sprouse had never served as a past president because he joined the Financial Accounting Standards Board in 1973 immediately after his year as president and had to resign from the Executive Committee. Thus, Sprouse feels that his year as director of publications was simply making up for his not getting to serve during 1973-74 [Sprouse, 1989].

There have been a few years when there were only nine individuals on the Executive Committee. As a result of a 1979 by-laws change, the office of editor of *The Accounting Review* was given the option of standing for reelection for additional one-year terms, but to avoid giving that individual an excessively long tenure on the Executive Committee, a provision was included that exempted the editor from the Executive Committee after three years. Since Stephen Zeff served five years as editor (1977-82), the 1980-81 and 1981-82 executive committees had only nine members. Similarly, Gary Sundem served four years as editor (1982-86), meaning that the 1985-86 Executive Committee had only nine members.

Exhibit 1-1
Members of Executive Committee: 1966-1991

Year	1966-1967
President	Lawrence L. Vance
Acad. VP	Clarence L. Dunn
Acad. VP	James B. Bower
Acad. VP	Gordon Shillinglaw
Prac. VP	Harold Nelson
Sec./Treas.	Joe Fritzemeyer
Research Dir.	Robert K. Jaedicke
Editor	Wendell Trumbull
Pres.-Elect	Frank S. Kaulback, Jr.
Past Pres.	Herbert E. Miller
Past Pres.	Robert Mautz

Year	1967-1968
President	Frank S. Kaulback, Jr.
Acad. VP	Harold Bierman, Jr.
Acad. VP	Peter A. Firmin
Acad. VP	J. E. Smythe
Prac. VP	Lawrence S. Dunham
Sec./Treas.	R. Lee Brummet
Research Dir.	Robert K. Jaedicke
Editor	Charles H. Griffin
Pres.-Elect	Sidney Davidson
Past Pres.	Lawrence L. Vance
Past Pres.	Herbert E. Miller

Year	1968-1969
President	Sidney Davidson
Acad. VP	Eldon S. Hendriksen
Acad. VP	Thomas F. Keller
Acad. VP	Edward S. Lynn
Prac. VP	John L. Carey
Sec./Treas.	R. Lee Brummet
Research Dir.	David Solomons
Editor	Charles H. Griffin
Pres.-Elect	Norton M. Bedford
Past Pres.	Frank S. Kaulback, Jr.
Past Pres.	Lawrence L. Vance

Year	1969-1970
President	Norton M. Bedford
Acad. VP	Robert T. Sprouse
Acad. VP	William J. Vatter
Prac. VP	Oscar Gellein
Sec./Treas	Catherine E. Miles
Research Dir.	David Solomons
Education Dir.	Stephen A. Zeff
Editor	Charles H. Griffin
Pres.-Elect	James Don Edwards
Past Pres.	Sidney Davidson
Past Pres.	Frank S. Kaulback, Jr.

Year	1970-1971
President	James Don Edwards
Acad. VP	Gerhard G. Mueller
Prac. VP	Robert E. Stevenson
Sec./Treas.	Catherine E. Miles
Research Dir.	Hector R. Anton
Education Dir.	Stephen Zeff
Editor	Eldon S. Hendriksen
Pres.-Elect	Charles T. Zlatkovich
Past Pres.	Norton M. Bedford
Past Pres.	Sidney Davidson

Year	1971-1972
President	Charles T. Zlatkovich
Acad. VP	Carl L. Nelson
Prac. VP	William L. Campfield
Sec./Treas.	Joseph A. Silvoso
Research Dir.	Hector R. Anton
Education Dir.	Harold Q. Langenderfer
Editor	Eldon S. Hendriksen
Pres.-Elect	Robert T. Sprouse
Past Pres.	James Don Edwards
Past Pres.	Norton M. Bedford

Year	1972-1973
President	Robert T. Sprouse
Acad. VP	Roland F. Salmonson
Prac. VP	Donald H. Cramer
Sec./Treas.	Joseph A. Silvoso
Research Dir.	Robert R. Sterling
Education Dir.	Harold Q. Langenderfer
Editor	Thomas F. Keller
Pres.-Elect	Robert N. Anthony
Past Pres.	Charles T. Zlatkovich
Past Pres.	James Don Edwards

Year	1973-1974
President	Robert N. Anthony
Acad. VP	Ray M. Powell
Prac. VP	Frank T. Weston
Sec./Treas.	Robert B. Sweeney
Research Dir.	Robert R. Sterling
Education Dir.	Doyle Z. Williams
Editor	Thomas F. Keller
Pres.-Elect	R. Lee Brummet
Past Pres.	Charles T. Zlatkovich
Past Pres.	James Don Edwards

Year	1974-1975
President	R. Lee Brummet
Acad. VP	Yuji Ijiri
Prac. VP	Wayne J. Albers
Sec./Treas	Robert B. Sweeney
Research Dir.	K. Fred Skousen
Education Dir.	Doyle Z. Williams
Editor	Thomas F. Keller
Pres.-Elect	Wilton T. Anderson
Past Pres.	Robert N. Anthony
Past Pres.	Charles T. Zlatkovich

Year	1975-1976
President	Wilton T. Anderson
Acad. VP	Donald H. Skadden
Prac. VP	Michael N. Chetkovich
Sec./Treas.	Floyd W. Windal
Research Dir.	K. Fred Skousen
Education Dir.	Robert L. Grinaker
Editor	Don T. DeCoster
Pres.-Elect	Charles T. Horngren
Past Pres.	R. Lee Brummet
Past Pres.	Robert N. Anthony

Year	1976-1977
President	Charles T. Horngren
Acad. VP	Alfred Rappaport
Prac. VP	R. M. Skinner
Sec./Treas.	Floyd W. Windal
Research Dir.	Thomas R. Dyckman
Education Dir.	Robert L. Grinaker
Editor	Don T. DeCoster
Pres.-Elect	David Solomons
Past Pres.	Wilton T. Anderson
Past Pres.	R. Lee Brummet

Year	1977-1978
President	David Solomons
Acad. VP	Thomas H. Williams
Prac. VP	William R. Gifford
Sec./Treas.	John K. Simmons
Research Dir.	Thomas R. Dyckman
Education Dir.	Leon E. Hay
Editor	Stephen A. Zeff
Pres.-Elect	Maurice Moonitz
Past Pres.	Charles T. Horngren
Past Pres.	Wilton T. Anderson

Year	1978-1979
President	Maurice Moonitz
Acad. VP	Kermit D. Larson
Prac. VP	Norman Auerbach
Sec./Treas.	John K. Simmons
Research Dir.	Eugene E. Comiskey
Education Dir.	Leon E. Hay
Editor	Stephen A. Zeff
Pres.-Elect	Donald H. Skadden
Past Pres.	David Solomons
Past Pres.	Charles T. Horngren

Year	1979-1980
President	Donald H. Skadden
Acad. VP	Sybil C. Mobley
Prac. VP	James Bulloch
Sec./Treas.	Alan P. Johnson
Research Dir.	Eugene E. Comiskey
Education Dir.	James M. Fremgen
Editor	Stephen A. Zeff
Pres.-Elect	Joseph A. Silvoso
Past Pres.	Maurice Moonitz
Past Pres.	David Solomons

Year	1980-1981
President	Joseph A. Silvoso
Acad. VP	John C. Burton
Prac. VP	Arthur R. Wyatt
Sec./Treas.	Alan P. Johnson
Research Dir.	A. Rashad Abdel-khalik
Education Dir.	James M. Fremgen
Pres.-Elect	Thomas R. Dyckman
Past Pres.	Donald H. Skadden
Past Pres.	Maurice Moonitz

Year	1981-1982
President	Thomas R. Dyckman
Acad. VP	John C. Burton
Acad. VP	William H. Beaver
Prac. VP	Arthur R. Wyatt
Sec./Treas.	Donald E. Kieso
Research Dir.	A. Rashad Abdel-khalik
Eduction Dir.	Thomas J. Burns
Pres.-Elect	Yuji Ijiri
Past Pres	Joseph A. Silvoso

Year	1982-1983
President	Yuji Ijiri
Acad. VP	Ray M. Sommerfeld
Acad. VP	William H. Beaver
Prac. VP	Robert K. Elliott
Sec./Treas.	Donald E. Kieso
Research Dir.	Theodore J. Mock
Education Dir.	Thomas J. Burns
Editor	Gary L. Sundem
Pres.-Elect	Harold Q. Langenderfer
Past Pres.	Thomas R. Dyckman

Year	1983-1984
President	Harold Q. Langenderfer
Acad. VP	Ray M. Sommerfeld
Acad. VP	Andrew D. Bailey, Jr.
Prac. VP	Robert K. Elliott
Sec./Treas.	Alvin A. Arens
Research Dir.	Theodore J. Mock
Education Dir.	Donald L. Madden
Editor	Gary L. Sundem
Pres.-Elect	Doyle Z. Williams
Past Pres.	Yuji Ijiri

Year	1984-1985
President	Doyle Z. Williams
Acad. VP	Charles G. Carpenter
Acad. VP	Andrew D. Bailey, Jr.
Prac. VP	Clarence Sampson
Sec./Treas.	Alvin A. Arens
Research Dir.	William R. Kinney, Jr.
Education Dir.	Donald L. Madden
Editor	Gary L. Sundem
Pres.-Elect	Stephen A. Zeff
Past Pres.	Harold Q. Langenderfer

Year	1985-1986
President	Stephen A. Zeff
Acad. VP	Charles G. Carpenter
Acad. VP	Robert S. Kaplan
Prac. VP	Clarence Sampson
Sec./Treas.	Jerry J. Weygandt
Research Dir.	William R. Kinney, Jr.
Education Dir.	Loren A. Nikolai
Pres.-Elect	Ray M. Sommerfeld
Past Pres.	Doyle Z. Williams

Year	1986-1987
President	Ray M. Sommerfeld
Acad. VP	William G. Shenkir
Acad. VP	Robert S. Kaplan
Prac. VP	David B. Pearson
Sec./Treas.	Jerry J. Weygandt
Research Dir.	Barry E. Cushing
Education Dir.	Loren A. Nikolai
Publications Dir.	Robert T. Sprouse
Pres.-Elect	William H. Beaver
Past Pres.	Stephen A. Zeff

Year	1987-1988
President	William H. Beaver
Acad. VP	William G. Shenkir
Acad. VP	Roger H. Hermanson
Prac. VP	David B. Pearson
Sec./Treas.	Frank R. Rayburn
Research Dir.	Barry E. Cushing
Education Dir.	Gary J. Previts
Publications Dir.	William W. Cooper
Pres.-Elect	Gerhard G. Mueller
Past Pres.	Ray M. Sommerfeld

Year	1988-1989
President	Gerhard G. Mueller
Acad. VP	Robert E. Jensen
Acad. VP	Roger H. Hermanson
Prac. VP	D. Gerald Searfoss
Sec./Treas.	Frank R. Rayburn
Research Dir.	Shyam Sunder
Education Dir.	Gary J. Previts
Publications Dir.	William W. Cooper
Pres.-Elect	John K. Simmons
Past Pres.	William H. Beaver

Year	1989-1990
President	John K. Simmons
Acad. VP	Robert E. Jensen
Acad. VP	Wanda A. Wallace
Prac. VP	D. Gerald Searfoss
Sec./Treas.	Joseph J. Schultz
Research Dir.	Shyam Sunder
Education Dir.	Corine T. Norgaard
Publications Dir.	Daniel W. Collins
Pres.-Elect	Alvin A. Arens
Past Pres.	Gerhard G. Mueller

Year	1990-1991
President	Alvin A. Arens
Acad. VP	Mark A. Wolfson
Acad. VP	Wanda A. Wallace
Prac. VP	David A. Wilson
Sec./Treas.	Joseph J. Schultz
Research Dir.	Nicholas Dopuch
Education Dir.	Corine T. Norgaard
Publications Dir.	Daniel W. Collins
Pres.-Elect	Arthur R. Wyatt
Past Pres.	John K. Simmons

American Accounting Association
Presidents
1966-1991

Lawrence L. Vance

Frank S. Kaulback, Jr.

Sidney Davidson

Norton Bedford

James Don Edwards

Charles T. Zlatkovich

Robert T. Sprouse

Robert N. Anthony

R. Lee Brummet

Wilton T. Anderson

Charles T. Horngren

David Solomons

Maurice Moonitz

Donald H. Skadden

Joseph A. Silvoso

Thomas R. Dyckman

Yuji Ijiri

Harold Q. Langenderfer

Doyle Z. Williams

Stephen A. Zeff

Ray M. Sommerfeld

William H. Beaver

Gerhard G. Mueller

John K. Simmons

Alvin A. Arens

Structure of Council and Nominating Committees

The true democratization of AAA began with the setting up of Council in 1978. The establishment of Council (originally called Advisory Council) was a move to assure that there would be broad membership involvement in the governance of the organization. A 1974 committee had recommended such a council, but the idea was abandoned at the time. A 1977-78 committee dealing with the roles of regions and sections, chaired by Thomas Williams, recommended an advisory council composed of section and regional leaders. That recommendation was approved by the Executive Committee in 1978 and the Advisory Council met for the first time at the Denver meeting. Initially a purely advisory group, Council became an official part of the governing structure with a 1980 bylaws amendment. Council, now composed of 30 members, is made up of representatives from all of the regions and sections, plus four members at large. In many respects, Council serves as a training ground for potential Executive Committee members of the future. Although Council provides an advisory role to the Executive Committee, probably its most important function is the electing of four members to the seven-person nominating committee.

The 1966 nominating committee was composed of five recent past presidents. As a result, there were criticisms that to be nominated a person had to be a member of the "good-ole-boy" network. Because of these criticisms, the Executive Committee passed a motion at the November 1974 meeting to change the composition of the nominating committee to the three immediate past presidents and three regional vice presidents elected by their peers (out of the eight regions then existing). A bylaws amendment to this effect was approved by the membership in August 1975. In 1976, an event occurred that brought about questions with respect to having the regional vice presidents represented as the non-insiders on the nominating committee; that event was the approval of special interest sections. Thus, at the March 1980 Executive Committee meeting, a bylaws amendment was approved that would change the composition of the nominating committee to seven members, including three past presidents and four persons elected by Council. These elected members are mostly members of Council, but they need not be. During the past decade about one-fourth of Council's selections have been nonmembers of Council.

Whereas the third quarter century began with a nominating committee composed totally of past presidents, the AAA has become more democratic with a committee that includes a majority of elected members.

Membership Trends

The overall membership of the AAA has stayed relatively stable or declined slightly over the past 25 years, but the composition of that membership has changed tremendously. In 1966, there were 10,762 members, of which about 70% were practitioners. Only 3,475 were academicians. In addition, there were 2,354 associate (student) members. By 1990, the total membership had declined to 9,303, of which 6,948, or 75%, were academicians. In other words, membership composition has changed from a 30-70 split to a 75-25 split. Exhibit 1-2 lists membership by year. Note that associate membership has declined along with practitioner membership. The associate membership began declining when *The Accounting Review* stopped publishing questions and answers from the CPA examination—a feature that was the most popular for the student members.

Emeritus membership, albeit small, has been increasing steadily since its establishment in 1974. An emeritus member is one who has been a member for at least 20 years and is at least 65 years of age. Emeritus members pay the same dues as associate members. There is also a life membership category for individuals who have been members for at least 50 years. Life members pay no dues. In 1991 there are 96 life members; however, many of these were purchased during the 1950s and 1960s when AAA sold life memberships for $100 or $125. Today, the only way to acquire life membership is to be a member for 50 years.

Foreign membership has been on the increase throughout the period. Today, there are 1,453 foreign academic members and 402 foreign practitioner members. Overall, 20% of the membership resides outside of the United States. Of these, Canada is the most heavily represented with 344 members; Japan is close behind at 298. There are 199 from Australia, 131 from Great Britain, and 112 from Korea.

Summary

Although membership has remained relatively stable, the past quarter century has been a period of dynamic growth for the AAA. The organization has become more democratized and much more reliant on academicians as its membership base. Programs have increased many fold over the past 25 years, and that is the subject of this book. The remaining chapters outline the educational and research programs which have been undertaken at the national, regional, and sectional levels of the AAA. Given the growing internationalization of the group, that topic is given special emphasis in Chapter 2.

References

Arens, Alvin A., "Celebration, Evaluation and Rededication," *Accounting Horizons*, December, 1990, pp. 88-96.

Gerhardt, Paul, Interview by Dale L. Flesher, March 8, 1991.

Minutes of the Executive Committee, American Accounting Association, 1965-1990.

Sprouse, Robert, Interview by Dale L. Flesher, December 12, 1989.

EXHIBIT 1-2

	Academic	Practitioner	Emeritus	Total	Associate Members
August, 1965	2,926	7,478	-	10,404	-
August, 1966	3,475	7,287	-	10,762	2,354
August, 1967	3,274	7,592	-	10,866	3,422
August, 1968	3,409	7,289	-	10,698	3,380
August, 1969	3,663	8,642	-	12,305	2,496
August, 1970	3,983	8,617	-	12,600	1,735
August, 1971	3,756	7,287	-	11,043	2,436
August, 1972	4,115	7,432	-	11,547	2,344
August, 1973	4,013	7,797	-	11,810	2,250
August, 1974*	3,950	6,749	62	10,761	1,631
August, 1975	4,159	6,543	90	10,792	1,648
August, 1976	4,552	5,957	114	10,623	1,663
August, 1977	4,731	5,626	130	10,487	1,663
August, 1978	5,359	5,505	158	11,022	1,567
August, 1979	5,691	5,048	150	10,889	1,316
August, 1980	5,967	5,001	175	11,143	1,037
August, 1981	6,218	4,875	181	11,274	1,413
August, 1982	6,162	4,653	176	10,991	1,550
August, 1983	6,270	4,311	197	10,778	1,478
August, 1984	6,285	3,992	209	10,486	1,991
August, 1985*	6,001	3,448	268	9,717	1,152
August, 1986	6,164	3,160	270	9,594	1,239
August, 1987	6,118	2,837	246	9,201	875
August, 1988	6,363	2,581	241	9,185	726
August, 1989	6,903	2,342	237	9,482	1,042
June, 1990	6,948	2,124	231	9,303	744

*Indicates first year of major dues increase.

U.S./Foreign Membership Breakdown in 1990

	USA	Foreign	% Foreign
Academic	5,495	1,453	20.9%
Practitioner	1,722	402	18.9%
Emeritus	220	11	4.8%
	7,437	1,866	20%

CHAPTER 2
INTERNATIONAL OUTREACH PROGRAMS

The American Accounting Association (AAA) is the largest organization of accounting educators in the world. Although the majority of members are from the United States, there are over 1,400 members from other countries. It may come as a surprise, but this large international membership makes the AAA the largest organization of non-American accounting educators in the world. To partially emphasize its international membership, the AAA recently adopted a new corporate symbol which appears on its letterhead and official publications. This symbol is a map of the world—and the United States is not even in the center of that map.

The Association's international emphasis is, however, of rather recent vintage. Although the AAA has had international members for decades, these foreign members received little attention from the Executive Committee until the past 25 years. In fact, Zeff's history of the AAA's first 50 years does not even mention the subject of the organization's role in international matters [Zeff, 1966]. Given Zeff's international orientation (the theme of his year as president was international accounting and culture), this omission of international subject matter illustrates the low emphasis that foreign members and international subjects received from the AAA in its first half century. This ignoring of international matters is not without reason. International accounting as a field of study is a product of the past quarter century, as is the growth in multinational business operations [Mueller, 1989]. Even *The Accounting Review* published very few articles on international subjects prior to 1966. There were 13 articles on a variety of international subjects authored by Mary E. Murphy (California State University in Los Angeles) between 1938 and 1961, and three articles by Gerhard Mueller between 1961 and 1965, but beyond these there were only a couple of personal experience articles written by individuals who had visited Europe [Previts and Committee, 1980].

With respect to international congresses, the AAA began to be represented with an official delegate to the congresses after they were renewed following World War II, even though the Association's Executive Committee did not feel disposed to spend any money for this purpose. What happened was that various people would receive an official status designation from the AAA president. This enabled the official delegate to sit at head tables, and to participate in "inside" decisions of the sponsors of the congresses regarding the selection of sites and other matters [Garner, May 10, 1990]. Paul Garner, Mary E. Murphy, Steve Zeff, George Scott, and Adolf Enthoven were all, at one time or another, designated as official AAA delegates at one or more conferences or international congresses. Despite this appearance of international activity, the AAA was doing nothing more than facilitating the travel of individuals who were already attending the conferences for personal reasons.

Things have now changed. Three Association annual meetings in the past 18 years have been held in Canada (1973, 1984, and 1990) and two have been held in the international city of Honolulu. There have been two Canadian officers and three presidents born outside of the United States. There is an official International Accounting Section with over 1,100 members. The AAA sponsors an international lecturer series that began in 1976 and maintains an International Professor Clearing House for faculty members who wish to teach in other countries. The American organization is affiliated with 11 other academic accounting organizations in other countries (Associate Organizations) and materials are sent gratis to libraries in several third-world countries. There have been numerous Association-sponsored publications on international accounting in recent years, and some have been translated into foreign languages such as Japanese and Spanish. Over the years there have been special international accounting meetings, and recently, international plenary session speakers at the annual meeting. The doctoral consortium now has foreign delegates and the Association sponsors an American doctoral student at the European Doctoral Colloquium. Finally, the AAA Executive Committee has held meetings outside of the United States on three occasions in recent years (the fall 1985 meeting was held in Bermuda, the fall 1987 meeting in Nassau, Bahamas, and the fall 1989 meeting in Amsterdam).

The overwhelming internationalization of the AAA has occurred quickly. Although this movement has required the approval of the various Executive Committees over the years, there have been four individuals, all past presidents of the AAA, who have been the primary movers in the internationalization of the organization. These four individuals were S. Paul Garner (president in 1951), David Solomons (1977-78), Stephen Zeff (1985-86), and Gerhard Mueller (1988-89). Surprisingly, these individuals' contributions to the movement did not all occur during their periods as president. Instead, all four individuals have made contributions throughout the past quarter century. The following sections outline the most important international outreach programs of the AAA.

International Accounting in the Curriculum

One of the first AAA initiatives in the international arena was the 1966 recommendation by the Committee on International Accounting that a course or seminar on international accounting be included in the accounting curricula of institutions of higher learning. It was argued that the need for such a course was "due to the ever-increasing expansion of international investment holdings of firms and individuals and the construction of subsidiary plants around the globe [Committee on International..., 1966, p. 2]." The committee was chaired by Gerhard Mueller (University of Washington). Other committee members were Louis M. Kessler (Alexander Grant & Company), Adolph Matz (University of Pennsylvania), J. E. Smyth (University of Toronto), and Paul Garner (University of Alabama).

Using arguments similar to the 1966 committee, the 1978 Education Committee of the International Accounting Section recommended not only an international accounting course, but an introduction to international accounting in the introductory principles course and a specialized major at the graduate level. The latter major would include a proficiency in one or more foreign languages and emphasis on cultures and political structures [Education Committee..., 1978, p. 3]. The 1978 committee was chaired by Norlin Rueschhoff (Notre Dame).

The International Lecturer Series

The next major international-oriented program was the International Lecturer Series which was inaugurated in 1976. This program was essentially the result of a donation to the AAA of $3,000 by Exxon Corporation for which there was no stipulation as to the use of the funds, except that it should be awarded to college professors or students for outstanding accomplishments. A committee chaired by S. Paul Garner met on November 7, 1974, and recommended that the money be used for a distinguished international lecturer program in which a distinguished lecturer from the United States would be sent on a lecture tour overseas and a distinguished foreign professor would be brought to the United States.

Paul Garner, who has now lectured in over 80 foreign countries, recently stated that the idea for the international lecturer series came about as a result of his many lectures in foreign nations. Often, he was asked about the possibility of having an exchange of professorships on an organized basis. Garner suggested such a program to 1974-75 President Lee Brummet and was placed in charge of the committee to develop the program [Garner, 1990].

The recommendation from the committee was essentially what was eventually adopted. It was suggested that each U.S. school on the lecture tour (ten were recommended) should contribute $200 to the AAA and the organization would contribute $1,500, making a total of $3,500 to pay to the distinguished lecturer. It was recommended that the program be financed by the donation from Exxon. There was speculation in the committee report that Exxon might want to provide future support for the program as well [Committee on International Professor..., 1974, p. 3].

The program was approved by the AAA Executive Committee when it met a week later [Minutes, Nov., 1974]. The committee was asked to provide more details of how the distinguished lecturers were to be selected. Final approval of the selection procedures came at the March 1975 meeting of the Executive Committee. The selection of the first distinguished lecturers was made in January 1976.

It was expected that $5,000 would be available to finance the two lecture series each year ($3,000 from the Exxon Controller's Grant and $200 each from ten U. S. host schools). This money was to be allocated between the two lecturers on the basis of the anticipated expenses which would be incurred. A four-member committee was to select the lecturers and the host institutions each year. Over the years, Exxon has stayed with the program and increased its contribution to $7,500 per year (in 1990). Schools in the USA who wish to serve as hosts now pay $450, plus the lodging and meal costs of the visitor while in the local community.

The program did not run too smoothly the first couple of years. The 1976 honorees were Robert R. Sterling of Rice University and Ray J. Chambers of the University of Sydney. Sterling found the Europeans quite hospitable, but felt that he was overscheduled. He found visiting 11 universities back to back exhausting, stating that he was usually scheduled from 7:00 a.m. to midnight. Despite the exhaustion of the trip, Sterling called the tour "one of the most worthwhile things" that he had ever done [Sterling, 1976, p. 2].

Chambers' report indicated that he, too, was overscheduled. He gave 58 addresses and seminars on 12 different subjects at ten universities. He also had 55 individual discussions with faculty and graduate students and 88 informal discussions at mealtimes. He concluded that although the hospitality extended him was excellent, there was no such thing as a "free lunch." He also complained that the $2,000 stipend he received was insufficient to cover his costs. In fact, his round-trip airfare alone amounted to $2,260. Other out-of-pocket costs exceeded $500. He could not even afford to bring his wife with him [Chambers, 1976].

The 1977 lecturers were Stephen A. Zeff, then of Tulane University, who had a successful nine-week tour of Latin America, and the late Edward Stamp of the University of Lancaster. Zeff delivered his 40 lectures (at ten institutions) in Spanish. To date, he is the only lecturer to deliver his lectures in a language other than English. Due to the staggering distances around Latin America, Zeff covered 20,000 miles in thirteen flights to visit eight countries.

Stamp complained from the beginning of the tour of being overscheduled. He then cancelled the end of the tour, much to the distress of the accounting administrators at the University of New Orleans and the State University of New York at Albany. Despite intensive arm-twisting from AAA president David Solomons, the trip to New Orleans was cancelled because it was "a city notorious for its heat and humidity" [Stamp, 1978, p. 3]. Because of the cancellations, Stamp's stipend from the AAA was reduced by $600 because of the $300 that had to be refunded to each of the schools not visited. Stamp was unhappy about this reduction in stipend. He also lambasted the AAA for giving Chambers and himself a smaller stipend than was given to the lecturers from the United States ($3,000 each for Sterling and Zeff; only $2,000 for Chambers and $2,400 after deductions for Stamp). Stamp's report even went so far as to complain because David Solomons did not write him a letter of thanks following the lecture tour (Solomons denied this allegation, but the timing of Solomons' letter cannot be determined). Interestingly, the Executive Committee later voted to reimburse Southern Methodist University for unstated extraordinary expenses incurred by that university in connection with Stamp's visit [Minutes, November, 1977]. Following these complaints, the stipend arrangement was changed so that each lecturer received a $1,000 honorarium plus reimbursement of all out-of-pocket expenses [Gerhardt, 1985, p.1].

There was no foreign lecturer for 1978 (the committee voted to invite someone from Latin America, but nothing came of that decision), and the 1979 lecturer, Baruch Lev, did not file a report. However, beginning in 1980 with Klaus Macharzina, the program stabilized and has gone smoothly most years since then. The only other problem year with the foreign lecturer was in 1985 when Hein Schreuder (University of Limburg), the youngest individual ever selected for the program, had to cancel because of marital problems involving a child custody battle [Schreuder, 1985, p. 1]. He had originally asked for a postponement of his lecture tour because of having moved to a new university. The Executive Committee granted this request for about 18 months, but finally had to withdraw the offer.

Herbert E. Miller, the 1981 lecturer from the United States, visited the Mediterranean rim countries. All went smoothly except for Tunisia, where there was no one to meet the distinguished lecturer. After several phone calls it was learned that his lecture had been cancelled due to student rioting and a change in University administration. Unfortunately, the new administration had informed neither Miller nor AAA of the cancellation. Miller had to spend a couple of days in Tunis anyway because there were no flights out other than the one upon which he was originally scheduled [Miller, 1981].

During Zeff's administration in 1985, the selection process was changed slightly so that selections of future lecturers were made 18 months in advance of the tour instead of only six months. Thus, at the August 1985 meeting, the executive committee selected both the 1986 and 1987 lecturers. The purpose of this change was to allow prospective lecturers to have more time to arrange their schedules to permit a six-week lecture tour. The first beneficiary of the extra planning time was 1987 lecturer Gerhard Mueller who has said that he never would have been able to accept the appointment without having a year and a half to plan for it [Zeff, 1989].

Below are listed all of the international lecturers from the United States and the countries they visited.

YEAR	INDIVIDUAL FROM USA	COUNTRIES VISITED
1976	Robert R. Sterling Rice University	Ireland, England, France, Scotland, Spain, Finland, Switzerland, Norway, Italy
1977	Stephen A. Zeff Tulane University	Argentina, Ecuador, Colombia, Chile, Mexico, Brazil, Venezuela, Peru
1978	Sidney Davidson University of Chicago	Japan, Hong Kong, Australia, New Zealand
1979	William H. Beaver Stanford University	England, Norway, Belgium, France, Germany, Switzerland
1980	George J. Benston University of Rochester	Argentina, Columbia, Chile, Mexico, Costa Rica, Brazil, Panama, Venezuela, Peru
1981	Herbert E. Miller University of Georgia	France, Tunisia, Egypt, Greece, Yugoslavia, Turkey, Israel, Spain, Italy
1982	George J. Staubus California-Berkeley	Hong Kong, Thailand, Japan, Singapore, Indonesia, South Korea, China
1983	Robert S. Kaplan Carnegie-Mellon	Ireland, England, Scotland, Netherlands, Switzerland, Sweden, Belgium, Austria
1984	David Solomons University of Pennsylvania	India, Bangladesh, Pakistan
1985	Yuji Ijiri Carnegie-Mellon	Australia, New Zealand, Japan
1986	William W. Cooper University of Texas	Costa Rica, Mexico, Peru
1987	Gerhard G. Mueller University of Washington	Nigeria, Zimbabwe, Kenya, Ghana, Zambia
1988	Alvin A. Arens Michigan State University	Peoples Republic of China, Singapore, Malaysia, Indonesia, Thailand
1989	William R. Kinney, Jr. University of Texas	France, Greece, Italy, Spain, Turkey, Yugoslavia
1990	Gary L. Sundem University of Washington	Australia, Japan, New Zealand
1991	Lawrence Revsine Northwestern University	Finland, Hungary, Poland, USSR

The foreign lecturers who have visited the USA and Canada have all been academics except one. Three of these individuals have come from the United Kingdom, three from Australia and two from West Germany. The foreign visitors and their affiliations are listed below.

YEAR INDIVIDUAL VISITING USA AND HIS HOME INSTITUTION

1976   Raymond J. Chambers, University of Sydney, Australia
1977   Edward Stamp, University of Lancaster, England
1978   NO VISITOR
1979   Baruch Lev, Tel-Aviv University, Israel
1980   Klaus Macharzina, University of Hohenheim, West Germany
1981   Anthony G. Hopwood, London Business School, England
1982   Hiroyuki Itami, Hitotsubashi University, Japan
1983   Robert J. Coleman, Commission of the European Communities,
       England and Belgium
1984   Bryan Carsberg, London School of Economics, England
1985   Hein Schreuder, University of Limburg, Netherlands, CANCELLED
1986   Murray Wells, University of Sydney, Australia
1987   A. G. Coenenberg, University of Augsburg, West Germany
1988   Andre Zund, Saint Gall Graduate School, Switzerland
1989   Abhulimen R. Anao, University of Benin, Nigeria
1990   Karel Van Hulle, Commission of the European Communities,
       and Catholic University of Leuven, Belgium
1991   Philip Brown, University of Western Australia

International Professor Clearing House

The AAA operates an international professor clearing house service for international professor exchange and placement. The Association has a file of schools which are interested in obtaining an accounting instructor on either an exchange or visiting basis. Currently, nearly 100 universities in the U.S. and Canada are on one list, while a similar number appear on a list of universities outside of the U.S. Canada appears on both lists. These lists are made available to interested members of the Association. All negotiations regarding an actual appointment or visit are then carried out between the individual and the host university.

The Clearing House program was the result of the same 1974 committee (chaired by Paul Garner) that recommended the establishment of the International Lecturer series [Committee on International Professor..., 1974, p. 2]. The program is one of the Association's lowest cost operations as the only requirements are a few hours of staff time per year.

International Accounting Publications

The Association's first publications in the area of international accounting were in the form of committee reports. For instance, the previously mentioned 1966 report of the Committee on International Accounting was issued in the form of a folded brochure. Later committee reports were published as a supplement to *The Accounting Review* from 1973 through 1977.

The 47-page committee report published in 1973 outlined the environment for international accounting. The committee, chaired by George M. Scott, then of the University of Texas, identified differences in modes of operations between domestic and multinational companies and the problems peculiar to operations abroad. The primary conclusion was that several aspects of multinational operations merit inclusion in the accounting curriculum ["Report of ...," 1973].

The next two reports of the Committee on International Accounting were substantially shorter than that published in 1973. Bertrand Horwitz (State University of New York at Binghamton) chaired the committee which produced the 1974 report. The committee's charge was to identify critical accounting issues facing those who teach or do research in the area of international accounting. The areas identified included performance evaluation in multinational enterprises, the impact of differences in national accounting principles and practices, foreign currency translation, accounting under central planning in the Soviet Union and East Europe (Chairman Horwitz had published a major study in this area), and the curriculum aspects of international accounting including the best vehicle for extending the subject to undergraduate students ["Report of ...," 1974].

Lee Seidler (New York University) chaired the next Committee on International Accounting and was able to keep the report to a mere five pages. The charge was limited in that the objective of the committee was to respond to the FASB on accounting for foreign currency translation ["Report of ...," 1975]. The 1976 supplement to *The Accounting Review* contained three international accounting committee reports. The 1974-75 Committee on International Accounting chaired by Kenneth S. Most (Florida International University) produced a 127-page report. The committee was given three charges, but the report covers only one of them—to study the impact of different national accounting principles and practices on financial statements by documenting the differences in a form suitable for use by researchers. Basically, the report analyzes the differences among six companies' financial statements—each company from a different nation ["Report of the American...," 1976].

The 1976 supplement also contained the report of a two-year Committee on Accounting in Developing Countries. George M. Scott chaired this committee which had the charge of recommending how the AAA should participate in efforts to improve accounting in developing countries. Unfortunately, the committee failed to meet its charge in that it found the biggest problem was the lack of qualified professors in developing countries and a lack of appreciation for those who were qualified. There was little the AAA could do to alleviate this problem ["Report of ...," 1976].

The third international committee report published in 1976 was a response to the International Accounting Standards Committee (IASC) on its exposure draft on consolidations and the use of the equity method of accounting. In 1977, three more responses to the IASC were published. Also published in 1977 was the report of the 1975-76 Committee on International Accounting Operations and Education, which was chaired by Adolf J. H. Enthoven (University of Texas at Dallas) ["Report of ...," 1977]. Other parts of this committee's report were later published under the sponsorship of Price Waterhouse with the title *Accounting Education and the Third World.*

The AAA issued its first major publication in the international area in 1970. That was Studies in Accounting Research No. 4, *Accounting Controls and the Soviet Economic Reforms of 1966,* authored by Bertrand Horwitz, then of Syracuse University. This monograph was an outgrowth of the Association's major research program which had begun in 1965. The study was commissioned by Robert K. Jaedicke (Stanford University) when he was Director of Research. The research advisory committee on the project consisted of Harold Bierman, Jr. (Cornell University), Carl L. Nelson (Columbia University), and David Solomons (University of Pennsylvania).

The next major international publication did not come until 1978 when *Accounting Education and the Third World* was published. This was an outgrowth of a committee report of the Association's Committee on International Accounting Operations and Education (1976-1978), chaired by Adolf Enthoven. David Solomons, then president of AAA, did extensive redrafting on the committee's report and contributed a foreword. The Price Waterhouse foundation financed publication. William Gifford of that firm was the AAA vice president that year and was involved in the project. The study was designed to improve accounting education and techniques in third-world countries. An accounting development framework was prepared.

Since 1978, there have been ten major international publications emanating from the Association; nine of these have come directly from the International Accounting Section, which was formed in 1976. These nine publications will be examined later in the discussion of the Section. The only recent international publication issued by the AAA that has not come out of the Section was the 1986 monograph entitled *Accounting and Culture,* edited by Barry E. Cushing. This volume consisted of the plenary session papers and discussants' comments from the 1986 annual meeting in New York City.

The Canadian Region

For a while, the AAA had a regional group based in a foreign country—Canada. A Canadian Region was approved by the Executive Committee in 1967. The organizational meeting and first conference of the region was held on June 21, 1967, at Carleton University in Ottawa, Ontario. The organizers of that meeting were Kenneth F. Byrd (McGill University), J. David Blazouske (Queens University), Brian E. Burke (University of British Columbia), W. Barry Coutts (University of Toronto), Howard B. Ripstein (Loyola College), Charles W. Schandl (University of Manitoba), and Michael Zin (University of Windsor). At the time, there were approximately 245 AAA members residing in Canada. By 1973, that number had increased to 310, over half of whom were practitioners [Minutes, August, 1973].

Kenneth F. Byrd served as the first chairperson of the Region, serving through three annual conferences. This long period of service was attributable to an oversight on the part of the Region's executive committee in not nominating new officers by the deadline specified by the AAA national office. In fact, Kenneth Byrd did not even attend the 1968 annual conference due to a lecture tour in Africa. Since the Canadians had failed to nominate new officers by the appropriate date, the 1967-68 officers continued to serve for an additional year [Byrd, 1969, p. 59]. Beginning with the 1969-70 committee, the members were for the most part selected from the region near where the next annual conference would be held in order to make it easy for the committee to meet without having to incur excessive travel costs.

The 1969 annual conference at York University was the first attended by an AAA president-elect, in the person of Norton Bedford. Bedford gave the usual official speech and then proceeded to give an eloquent oratory on the future of accountancy. His emphasis was on mathematics, behavioral science, the use of fifth generation computers between 1980 and 1985, and the certainty of great advances in socioeconomic accounting [Byrd, 1969, p. 60-61].

Due to the great distances covered by the Canadian Region, the annual meetings were never highly attended. In fact, the AAA Executive Committee at times discussed the idea of terminating the Region because most members would actually be closer to a regional meeting in the USA than they would be to their own Canadian Regional meeting. Even in the later years, the Canadian Regional meetings drew only about 100 attendees.

To help solve the problem of attendance, the Region met the day before the annual meeting of the Association of Canadian Schools of Business (an association of academics of all business disciplines except accounting). In 1973 this latter association became the Canadian Association of Administrative Sciences (CAAS) and added Accounting and Management Information Systems as a division. The Cana-

dian Region held a separate meeting each year through 1973. Beginning in 1974, Professor Barry E. Hicks (University of Western Ontario), as chairperson of the Canadian Region and chairperson of the Accounting and Management Information Systems Division of CAAS, held the first joint meeting of the two organizations. These joint meetings with CAAS continued until and including 1976. In the 1980's CAAS changed its name to the Administrative Sciences Association of Canada.

In 1977, before tax changes were implemented that would hinder the raising of funds from accounting firms, the Canadian academic accountants formed their own organization, called the Canadian Academic Accounting Association (CAAA). The AAA Canadian Region switched in 1977 from meeting jointly with the CAAS to meeting jointly with the CAAA. Meeting jointly was approved by the AAA Executive Committee at the August 1976 meeting in Atlanta. In reality, the AAA Region and the CAAA were both the same organization. In fact, the same person served as president of the CAAA and as the AAA Regional vice president.

Another explanation for the break-up of the AAA Region was that during the late 1970's and early 1980's, the Canadians experienced a period of nationalism that fueled the march toward a separate Canadian association that was not an affiliate of a U. S. group. It was felt that a Canadian chartered group would be able to more easily obtain funds from Canadian accounting firms and foundations.

Finally, in February 1984, the CAAA executive committee decided to dissolve the Canadian Region, which, it was noted, had been inactive for several years [Gibbins, 1984, p. 1]. The AAA Executive Committee officially terminated the Canadian Region at the August 1984 meeting.

The table on the next page lists the locations of the annual Canadian Region meetings and the Regional chairperson or vice president. It is notable that two of the later Regional chairpersons (Alister K. Mason and Gilles Chevalier) were not educators (however, they both held Ph.Ds and were former professors). The listing was compiled from letters received from several former regional chairpersons, with most of the information being provided by Barry Hicks [1990]. The list varies somewhat from the data available in the Association's files in Sarasota due to the fact that the AAA's national office was not always in full communication with the Canadian group. As a result, archival materials are scant. The annual meeting was rarely announced in *Accounting Education News* because the location and dates were not known until the last minute (at least not in Sarasota). This was in contrast to the actions of other regions where annual meetings were scheduled years in advance.

One major contribution of the Canadian Regional Group was that the Proceedings of the annual meetings were published in some years, beginning as early as 1969 and 1970. Brian Burke of the University of British Columbia was responsible for the publication. Unfortunately, there is nothing to indicate that the national office

OFFICERS AND LOCATIONS OF AAA CANADIAN REGION

Year	Location	Regional VP
1967	Carleton University, Ottawa, Ontario	Kenneth F. Byrd, McGill University
1968	University of Calgary, Calgary, Alberta	Kenneth F. Byrd, McGill University
1969	York University, Toronto	Kenneth F. Byrd, McGill University
1970	University of Manitoba, Winnipeg, Man.	Charles Schandl, Dalhousie University
1971	Memorial University, St. Johns, Newfoun.	Brian E. Burke, University of British Columbia
1972	McGill University, Montreal, Quebec	B. R. Howson, McGill University
1973	Queen's University, Kingston, Ontario	R. G. Laybourn, McGill University
1974	University of Toronto, Toronto	Barry E. Hicks, University of Western Ontario
1975	University of Alberta, Edmonton, Alberta	Leonard G. Eckel, McMaster University
1976	Laval University, Quebec City, Quebec	Leonard G. Eckel, McMaster University
1977	University of New Brunswick, Federicton	Daniel L. McDonald, Simon Fraser University
1978	University of Western Ontario, London, ON	L. S. (Al) Rosen, York University
1979	University of Saskatchewan, Saskatoon	John Waterhouse, University of Alberta
1980	Universite du Quebec A Montreal, Montreal	David Blazouske, University of Manitoba
1981	Dalhousie University, Halifax, Nova Scotia	Alister K. Mason, Deloitte, Haskins, & Sells
1982	University of Ottawa, Ottawa, Ontario	W. John Brennan, University of Saskatchewan
1983	University of British Col., Vancouver, BC	Gilles Chevalier, Touche Ross
1984	University of Guelph, Guelph, Ontario	Michael Gibbins, University of British Columbia

knew of the publication. In fact, much ado was made in 1974 when the Southeast Region published its Proceedings of the Memphis meeting. It was widely believed that the Southeast had been the first region to publish Proceedings. The 1975 and 1976 Canadian AAA Proceedings were published in microfiche format by the Canadian Association of Administrative Sciences, with whom the Region met jointly, and in hard copy by the Region. The 1977 and 1978 Proceedings were published together in one volume under the joint sponsorship of the Region and the CAAA. From 1980 through 1984, the Proceedings were published by the CAAA without mention of the AAA Region. Apparently, the facade of joint conferences was dropped in 1979 or 1980 [Hicks, 1990].

Other International Programs

Associate Organizations

The AAA initiated an outreach program toward foreign organizations of academic accountants during the presidency of Stephen Zeff in 1985-86. The aim of the program was to strengthen the ties with other academic accounting bodies and to promote activities of mutual interest. The idea for such a program came out of a 1985 Committee on Relations With Other Organizations chaired by Yuji Ijiri. It was

probably appropriate that the responsibility for initiating the program was turned over to Zeff since he was familiar with most of the organizations that qualified for the program [Zeff, 1989].

The benefits to these organizations of being AAA associates are:

(1) To nominate a doctoral student to be eligible to attend the AAA's annual Doctoral Consortium (however, there will be a maximum of four foreign participants each year);

(2) To have the registration fee of the organization's president waived at the AAA annual meeting.

(3) To have their annual conference and publications announced in *Accounting Education News.*

(4) To have the table of contents of the Associate Organizations' research journals carried in *The Accounting Review.*

(5) To have both the president and depositary library receive complimentary copies of AAA journals and other publications.

The AAA asks for reciprocal privileges, where applicable, from the Associate Organizations [Zeff, May, 1986, p. 7].

Initially, there were eight associate organizations. Today, there are the eleven listed below:

Accounting Association of Australia and New Zealand
British Accounting Association
Canadian Academic Accounting Association
European Accounting Association
French Accounting Association
Indian Accounting Association
International Association for Accounting Education and Research
Japan Accounting Association
Korean Accounting Association
Nigerian Accounting Teachers' Association
Southern African Society of University Teachers of Accounting

Another program involving international institutions was Zeff's initiative to send AAA publications to libraries at tertiary institutions in third-world countries. The idea originally emanated from a 1984-85 Committee on Relations With Developing Countries, chaired by Raj Aggarwal (University of Toledo). A committee of the International Accounting Section, with the cooperation of the U.S. State Department, ultimately suggested the names and addresses of 57 institutions of higher learning in 23 countries who were worthy of such aid from the Association [Zeff, November, 1985, p. 5].

KMG/Main Hurdman International Conference

One of Zeff's major contributions to the international movement was the mounting of an international conference on standard setting, held in Princeton, New Jersey, just before the 1986 annual meeting. This was the Association's first conference not dedicated to accounting research or education. The intention was to provide guidance to standard setters in both developed and developing countries in charting the future of their programs. The conference enabled representatives from standard setting organizations to review their progress to date, discuss common problems, and examine the role of research in standard setting as well as the likely costs and benefits of articulating a conceptual framework. The conference was coordinated by past AAA president David Solomons [Zeff, October, 1985, p. 3].

The biggest problem in mounting the conference was obtaining funding. Zeff received approval from the Executive Committee for the conference in November, 1984. Initially, Zeff approached a Big-8 CPA firm. After a delay of many months, the firm declined to participate. At that point, Zeff went to KMG in Amsterdam with the argument that an international conference sponsored by that firm would heighten the awareness of the name of that firm's U. S. affiliate—Main Hurdman. KMG offered to cover the costs of the conference up to $75,000. Ultimately, the total cost was only about $55,000 [Zeff, 1989]. Most of the conference participants were able to stay for the AAA annual meeting, a factor which resulted in even greater international visibility for the AAA.

Attendance at the conference was by invitation only. This resulted in several researchers being disappointed in not being able to attend the conference. Zeff and Solomons felt the attendance should be limited to insure that the standard setters would be able to speak freely with respect to their roles. The idea was to provide a small-group setting in which the participants could come to know each other and engage in a useful dialogue on matters of mutual interest. Representatives came from 23 countries [Zeff, May, 1986, p.7; and Solomons, November, 1986, p. 5].

Books for Black Africa Project

One of the most successful of the recent international outreach programs was the 1988-89 efforts to acquire accounting books to be sent to universities in underdeveloped African countries. This program sprouted from Gerhard Mueller's tour as the Association's 1987 Distinguished International Lecturer to eight African universities. Mueller became aware of the shortage of books in those countries and organized a campaign to obtain recently published accounting textbooks to be sent to African institutions. The campaign was quite successful as the Association's office in Sarasota was overwhelmed with contributions of hundreds of books. Several

Beta Alpha Psi chapters assisted in the book collection effort. The books were sorted by title and mailed off in late 1989 ["Report...," January, 1989, p. 3].

A similar project had been recommended to the International Accounting Section in 1981 by Yaw Mensah, then at Indiana University. Mensah's recommendation was that books be collected and distributed to universities in developing countries. The Section officers declined to get involved in such an undertaking because of a belief that the Section had neither the time nor the facilities to handle such a project [Sinning, 1981]. There have been other attempts to send books to foreign institutions, but these have had no more Association sponsorship than a mention in *Accounting Education News.*

International Speakers at the Annual Meeting

Over the years, there have been international professors appear on the concurrent sessions of the AAA annual meeting. Rarely, however, have non-U.S. speakers appeared on the plenary sessions at the Association's annual meeting. An Australian speaker, Russell Mathews of the University of Adelaide, did appear on the program of the 1959 meeting in Colorado speaking on inflation accounting, but such instances of foreign participation were rare. There were Canadian plenary speakers at the Quebec meeting in 1973. After that, Edward Stamp spoke at the Portland meeting in 1977. The next international plenary speakers at a USA annual meeting who were not a part of the Lecturer Series were Geert Hofstede (University of Limburg, Hein Schreuder (University of Limburg), and Anthony G. Hopwood (University of London) who spoke at the 1986 meeting in New York ["Annual...," June, 1986, pp. 5-9]. The plenary session papers and discussants' comments were published in a monograph entitled *Accounting and Culture* [Cushing, 1986]. In addition, Julian Paleson (Grant Thornton, Brussels) was a luncheon speaker ["A Delicious..." 1986, p. 5].

The theme of the 1989 Hawaii annual meeting, presided over by President Gerhard Mueller, was "Meeting International Challenges." The opening ceremony featured greetings from eight of the AAA's Associate Organizations ["Hawaiian..., October, 1989, p. 9]. The keynote address of that meeting was made by Charles E. Young, the chancellor of UCLA, who spoke on the subject of "The International Imperative of the Modern University." Featured international speakers included Norishige Hasegawa (Japan) and Sidney J. Gray (University of Glasgow). Some of the AAA sections followed President Mueller's international theme by emphasizing international speakers for their section meetings in Hawaii. The Auditing Section, for example, hosted Jack W. Flamson (Price Waterhouse), who spoke on "International Auditing: Managing the Differences" [Flamson, Fall, 1989, p. 2ff].

Since 1968 at San Diego, the annual meeting had been the focus for an Association-sponsored reception for foreign registrants [Minutes, March 1968, p. 8].

Despite criticisms of the cost from the Executive Committee [Minutes, December 1968, p. 11, and April 1970, p. 13], the receptions continued to be held through 1976, but none was held in 1977. In 1978, Anita Tyra, then chairperson of the International Accounting Section, requested that the reception be reinstated. The 1978 reception was a Section-sponsored activity, but was financed by AAA funds. Tyra, however, got the idea for reinstating the reception from Association president David Solomons who wanted to see the reception reinstated. Knowing of the criticisms of the idea that had emerged over the years, Solomons apparently was reluctant to recommend the reception to the Executive Committee himself, so he urged Tyra to make the request. Solomons then approved her request [Tyra, March 13, 1978].

The International Accounting Section

In 1975, the AAA initiated a change in policy that permitted the establishment of special interest sections [March, 1975, p. 10]. One of the first group of sections founded (August 23, 1976) was the International Accounting Section. Since that time, the International Accounting Section has played a major role in the Association's international outreach programs. The initial officers of the new section were Hanns-Martin W. Schoenfeld (University of Illinois), Chairman, Anita Tyra (Bellevue Community College), vice chairman, and Irving Fantl (Florida International University), secretary-treasurer. The organizers were careful to select the initial officers from the two primary centers of influence in the international accounting arena—the University of Illinois (Schoenfeld) and the University of Washington (Tyra)—and from individuals independent of the two centers (Fantl). This balance of officers has continued to be observed over the years. The officers were aided by an advisory board of 15 scholars and practitioners from several countries. S. Paul Garner was the first board chairman [Schoenfeld, 1977, p. 1]. It so happens that Garner was the first dues-paying member of the Section. A letter in the Association's files from Paul Gerhardt to Paul Garner thanks Garner for his $50 contribution to the Section and informs him that his payment "holds the distinction of being the first revenue item, dues or contribution, which has been received by one of our new sections" [Gerhardt, November 12, 1976].

Despite Gerhardt's note to Paul Garner informing him that he could lay claim to being the first dues-paying member, the first true members on the Section's membership petition were Gerhard Mueller and Kenneth B. Berg (in that order) of the University of Washington. In 1976, petitions for the creation of a section had to contain 100 names. It was Mueller and Berg who took the initiative of obtaining the 100 names. The petition forms were on the stationery of the International Accounting Studies Institute at the University of Washington. A total of 120 names were submitted with the petition Two other individuals, Kenneth Most and Irving Fantl of Florida International University, tried at the same time to start a petition for an international section, but were told by Paul Gerhardt that Mueller had already begun the process [Gerhardt, June 3, 1976].

An organizational meeting was held at the 1976 annual meeting in Atlanta. Sixty persons were in attendance at that first meeting. The Section's bylaws were passed the following year at the Portland annual meeting [Garner, 1985, p. 4].

Initially, the International Accounting Section's dues were $10 per year, the highest of any of the initial five sections other than Public Sector, which also had dues of $10. The lowest dues of any section was $5. Despite the high dues, the Section grew quickly. By the end of 1977, membership stood at 352 members, 60 percent of whom were from overseas. The initial growth of the Section can be attributed to the extensive mailing lists that had already been compiled by the Universities of Illinois and Washington and by Irving Fantl, who had published an international accounting newsletter. Membership had increased to 743 by the end of 1980 and then stayed in that neighborhood for the next six years. There was another major increase in membership in 1987 when the number grew to 921. By the end of 1988, that number had increased to 1,101. The 1980's ended with membership in the Section at 1,155. Approximately one-third of the members are from outside of the USA with 53 countries being represented. The International Accounting Section is the fourth largest section behind Auditing (1,620 members), Management Accounting (1,524 members), and the American Taxation Association (1,213 members).

The Section's finances grew commensurate with the growth in membership. The Section closed out its 1977 fiscal year with a cash balance of $3,002. The cash position varied from this figure by only about $2,000 over the next six years. The balance was down to $2,349 at the end of 1983. Subsequent years were good to the Section. Cash on hand amounted to $10,622 by the end of 1984 (expenses for that year were only $1,483 against revenues of $9,756). By the end of fiscal year 1989, the cash balance had grown to $27,228. Part of the reason for the increase in funds was that the officers were stockpiling cash in order to establish a Section journal.

The official objectives of the International Accounting Section are to:

1. Encourage, support and promote interest in all aspects of international accounting both in the USA and throughout the world.
2. Provide a means of communication among those interested in international accounting through regional and national meetings, publication of a membership roster and newsletter, preparation of other publications relevant to international accounting, collection and dissemination of information about international accounting courses, encourage international faculty exchange, and support the international liaison activities conducted by the AAA administrative office or other committees.
3. Encourage international accounting research and provide a forum for exchange of research findings through appropriate programs at annual Section meetings

and regional meetings and by monitoring important international accounting events and foreign research.

4. Facilitate, as far as feasible, special research, teaching, or information needs of Section members as they arise [1989-90 Directory, 1989, p. 32].

These objectives were supplemented in 1989 with the passage of a Section mission statement. According to that document, the mission of the Section is to take a leadership role in teaching and research in international business and accounting. To that end, the Section shall (1) represent the academic interest of its members in the AAA and other domestic, foreign, and international professional organizations, (2) facilitate research and exchange of ideas in the international dimensions of accounting, (3) promote teaching of the international dimensions of accounting and internationalizing business school curricula, and (4) sponsor conferences, publications and other forms of disseminating knowledge about the international dimensions of accounting ["AAA—International...," 1989, p. 8].

Section Publications

The first publication of the International Accounting Section was a newsletter entitled *Forum.* The first issue came out in early 1977 under the editorship of Irving Fantl. Fantl had earlier (January 1973 through Spring 1974) published his own newsletter entitled *Forum on International Accounting* while he was a faculty member at Rider College in New Jersey. There was then an interruption while he completed his dissertation. As the secretary-treasurer of the new Section, Fantl took the opportunity to reestablish his old newsletter, but under the auspices of the Section. Fantl remained as editor for five issues [Gibson, 1986, p. 5]. Initially, the *Forum* was published three times per year, but became a quarterly in 1980.

The contents of the *Forum* consists primarily of news items regarding the Section, its officers, and relevant meetings of other organizations. There is always a chairperson's message and usually committee reports and minutes of executive board meetings. Occasionally, there is a short article or other item of interest. For example, the January 1980 issue contained a list, supposedly complete, of all dissertations (83 of them) ever written in the field of international accounting. The list was compiled by Abdel Agami and Charles Fazzi. Subsequent issues of the *Forum* kept the list up to date.

Another Section publication was the membership roster which has been issued every couple of years. The first roster came out in 1977 and cost the Section about $1,000, as did the 1979 version. The cost increased to $2,000 for the 1980 volume, while the 1983 edition cost more than $3,000.

To date, the International Accounting Section has published nine monographs, the first of which came out in 1979. The first major publication of the Section was *Notable Contributions to the Periodical International Accounting Literature—1975-78*. This project was started in 1978 under the interest and efforts of Anita Tyra, who was then chairperson of the Section. Printing costs came to $6,588.

In 1980, the Section published *Eighty-eight International Accounting Problems in Rank Order of Importance—A DELPHI Evaluation*, by George M. Scott and Pontus Troberg. This 115-page monograph provided the first systematic study identifying, evaluating, and assessing research methodology for resolving major international accounting problems. The study was commissioned by Lee Radebaugh, while chairperson. The study presents a cogent model for examining international accounting problems. Publication costs came to $4,675.

The third in the Section's series of publications was the 1982 monograph by Jane O. Burns, Michael A. Diamond, and Helen Morsicato Gernon entitled *The International Accounting and Tax Researchers' Publication Guide*. Listed were 88 journals (54 of them in the USA) which publish international articles. Printing costs totaled $2,797.

The fourth of the Section's monographs also came out in 1982 under the title of *A Compendium of Research on Information and Accounting for Managerial Decision and Control in Japan*, edited by Seiichi Sato, Kyosuke Sakate, Gerhard Mueller, and Lee H. Radebaugh. This was a volume of readings by Japanese scholars. The articles were first written in Japanese and then translated into English. The idea for the volume was promoted by Professor Sakate while visiting the University of Washington in December 1977. Gerhard Mueller's preface states:

> Twenty years ago, the field of international accounting did not exist. Since then it has solidly established itself as a curriculum component in accounting education at colleges and universities,.... This publication is a giant step toward the age-old dream of a functioning international community of scholars. Publication of the present compendium demonstrates that recognized scholars, working in one particular language and cultural setting, can transfer their work to those similarly engaged in other settings.

The volume cost $4,882 to print and another $1,850 was paid to the Japanese professors who served as editors and translators.

In 1983, another monograph was published. This was *Annotated International Accounting Bibliography, 1972-1982*, by Abdel M. Agami and Felix P. Kollaritsch. This listing included nearly 1,000 articles published in 35 periodicals. It was published at a cost of $4,195.

Belverd E. Needles edited a volume in 1985 entitled *Comparative International Auditing Standards*. This monograph provided an introduction to auditing standards in various countries of the world and examined the possibilities for harmonization in ten countries. This monograph and the one listed below on taxation were both the result of joint committees formed with other AAA sections during the administration of Section chairperson Konrad Kubin of Virginia Tech University.

The next Section publication was *Comparative International Taxation*, edited by Kathleen E. Sinning, which came out in 1986. This volume was the result of the efforts of a Joint International Accounting Section—American Taxation Association Research Committee on International Taxation. The committee's charge was to examine eleven countries and:

1. Document the sources of tax revenue and their relative importance in selected countries.
2. Document the tax treatment of foreign source income by these countries.
3. Document the tax treatment of foreigners and foreign corporations earning income in these countries.
4. Identify similarities and differences in the above.

In 1987, the Section published *Proceedings of the V International Accounting Congress on Accounting Education*, edited by Juan M. Rivera and Konrad W. Kubin. When the editors discovered that the proceedings of this Mexican conference had not been published, they approached the Section officers with a request that the Section underwrite the publication costs.

Also in 1987, the Section published *Cases in International Accounting*, edited by James A. Schweikart. The main volume consisted of eleven cases suitable for use in an international accounting course and was accompanied by a smaller volume subtitled *Teaching Notes and Solutions*.

An International Section Journal

Despite being one of the earliest and largest of the AAA sections, the International Accounting Section has never had its own journal. This lacking has not been due to oversight or a shortage of effort. The subject of whether to publish a journal has been one of the most controversial issues that the Section officers have faced over the years.

The subject of a journal was first discussed at the March, 1983, Section executive committee meeting. It was estimated that the cost to publish a journal would be about $4,000 per issue. The issue was turned over to a Publication Committee for investigation. The committee decided that the costs would be excessive, given that at the time the Section's treasury was carrying a balance of just over $2,000 [Minutes of Executive..., October, 1983, p. 4 and 8]. Konrad Kubin, 1983-84 chairman, continued to pursue the issue. He noted in his "Chairperson's Message" that

Vernon Zimmerman had pointed out the fact that the University of Illinois's *International Journal of Accounting Education* received more worthwhile articles than it could accommodate [Kubin, November, 1983, p. 1]. Actually, Kubin had gone to Zimmerman with the idea of the Section cosponsoring the Illinois journal, but Zimmerman was apparently not interested in the offer. Kubin hoped for a University to cosponsor the journal in order to share costs. In March, 1984, it was noted that no interested schools had been found [Kubin, March, 1984, p. 2]. Later in 1984, a committee consisting of Konrad Kubin, Ray Sommerfeld and Lee Radebaugh was appointed to plan the early implementation of a new international accounting journal. The first issue was planned for late 1985 or early 1986. Kubin's hope of having a cosponsor for the journal was answered by Ray Sommerfeld who offered to get the University of Texas to provide financial support through Sommerfeld's Bayless/Rauscher Pierce Refsnes Chair. The Texas support was to be in the amount of $4,000 the first year, $3,000 the second year, and $2,000 in the third year [Holzer, 1984, p. 1]. Sommerfeld asked only that the University of Texas be listed as a cosponsor of the journal.

A Name-That-Journal Contest was held and the winner was Robert Raymond of the University of Nebraska, who submitted the title *International Accounting Review* ["Minutes of Executive..., June, 1985, p. 6]. That title, however, had to be changed due to opposition from the AAA Executive Committee due to possible confusion of the Section journal with *The Accounting Review* published by the parent organization. Thus, at the August, 1985, meeting, Kubin stated that three titles were acceptable. The journal was to be a semiannual publication. Costs were anticipated to be $9,000 per year in excess of advertising and subscriptions. With the support of the University of Texas and the significant build-up of Section Funds, the journal could be started without increasing dues ["Minutes of Executive...," October, 1985, p. 2]. The Section's cash balance had by then accumulated to over $14,000.

Momentum faltered, however, as H. Peter Holzer, the 1984-85 chairperson tabled the initiative during his administration, probably because of his closeness to the Illinois journal (he was on the review board). His successor, Helen Gernon, the 1985-86 chairperson, noted in her January, 1986, "Chairperson's Message" that the establishment of a Section journal was placed on hold until an agreeable editor could be identified [Gernon, January, 1986, p. 1]. She made a similar comment at the August, 1986, annual meeting. Ultimately, Ray Sommerfeld had to withdraw his offer of support.[1]

---

[1]Coincidentally, Sommerfeld was accused a couple of years later when he was AAA president of creating a tension between the Section and the international activities of the parent organization. Belverd Needles, 1987-88 Section chairperson, stated that symptomatic of this schism was the failure of Sommerfeld to invite Section officers to the reception for international attendees at the 1987 Cincinnati convention (Needles, Aug. 20, 1987). Sommerfeld apologized to Needles for the oversight in not inviting Section officers (Sommerfeld, Sept. 2, 1987).

Gernon has been accused of abandoning the issue because of political reasons going back to the old "centers of influence" controversy. Basically, she and the executive committee could not agree on an editor because the Washington group wanted one of their people as did the Illinois and independent groups.

There was silence on the topic of a journal until early 1988, when 1987-88 president Belverd Needles made an impassioned plea for the establishment of a Section journal. Needles argued that the Section should be the primary organization for the advancement and dissemination of research in the field of international accounting. He felt that this could not be accomplished without publishing a high quality journal. He argued that a high quality research journal was the mark of an academic discipline. "It is a distinguishing characteristic from a fraternal group with an interest in a particular area" [Needles, Winter, 1988, p. 1]. Needles summarized the arguments against a Section journal as being: (1) there are already sufficient outlets for international research, (2) there are insufficient articles representing high quality research to use in a new journal, and (3) the cost is too high. Needles countered these arguments with several of his own. First, he postulated that the lack of a strong refereed journal from an international organization like the International Accounting Section is a primary reason why there is not more high quality research being undertaken. With a journal editor who can work with a board of reviewers, promising research can be nourished through several revisions. He observed that the lack of quality research may be due solely to the lack of a nurturing environment for such research [Needles, Summer, 1988, pp 1-2].

Unfortunately, Needles' term as chairperson ended before a journal could be established. For his immediate successors, Kenneth Most and Kathleen Bindon, the establishment of a journal was not a high priority. In fact, shortly after Bindon took office, she stated to the AAA Executive Committee that the Section would not pursue starting a journal while she was chairperson, even though the funds had been accumulated for that purpose [Bindon, September 26, 1989].

Annual Breakfast/Lunches

In 1984, the Section began holding annual breakfasts or lunches in conjunction with the AAA annual meeting. The speaker for the 1984 Toronto breakfast was Phillip P. Aspinall of Coopers & Lybrand. At the time, Aspinall was president of the Canadian Institute of Chartered Accountants. At the request of Felix "Phil" Pomeranz, Coopers & Lybrand underwrote the cost of breakfast. The 1985 breakfast speaker in Reno was G. B. Mitchell, former secretary general of the International Accounting Standards Committee and then Technical Director of the Institute of Chartered Accountants in England and Wales. For the 1987 breakfast in Cincinnati, David

Cairns was the speaker. Cairns was the Secretary General of the International Accounting Standards Committee. The CPA firm of Arthur Young & Company sponsored the breakfast.

In 1988, the annual get together was changed to a luncheon. The speaker in Orlando was Stephen D. Harlan, Vice President International of Peat Marwick Main & Co. There was no sponsor and members had to pay $17 per person for the luncheon. The luncheon speaker for the 1989 Hawaii meeting, again unsponsored, was FASB chairman Dennis R. Beresford, who spoke on the role of the FASB in the internationalization of accounting standards.

Dissertation Awards

In 1984, the International Section began presenting an annual dissertation award. The award was first announced in November 1983 and the first award period was to cover dissertations written during the two years prior to January 1, 1984. There were 15 applicants for the first award, which was won by Trevor S. Harris (Columbia University). His dissertation was written at the University of Washington under Gerhard Mueller. In 1985, no award was granted due to the fact that there were only two applications. The two applications were carried forward to the following year. The application period for the 1985 award covered only one year while the 1984 period had covered the two preceding years. Other winners of the Dissertation Award and their universities are listed below along with the titles and the dissertation chairperson. Two awards were given in 1989.

YEAR	WINNER, TITLE, PROFESSOR	DOCTORAL UNIVERSITY
1984	Trevor S. Harris "Consequences of Economic Concepts of Foreign Exchange for Financial Accounting," chaired by Gerhard Mueller.	University of Washington
1985	NO AWARD	
1986	Betty C. Brown	Virginia Tech University
1987	Shahrokh M. Saudagaran "The Effect of Financial Reporting Requirements on the Decision to List on Foreign Stock Exchanges," chaired by Gerhard Mueller.	University of Washington
1988	David Sharp "Control Systems and Decision-Making in Multinational Firms: Price Management Under Floating Exchange Rates," chaired by Donald R. Lessard.	M.I.T.

1989	Terry L. Conover "An Empirical Investigation of the Effects of the Accounting Treatment of Foreign Currency Translation on Management Actions in Multinational Firms," chaired by Wanda Wallace.	Texas A & M University
1989	Zelma Rebmann-Huber "The Influence of Various Groups on Accounting Standard Setting in Sixteen Developed Countries: Model and Empirical Investigations," chaired by Gerhard Mueller.	University of Washington
1990	F. Norman Shiue "A Positive Theory of Accounting Standards Determination: The Case of Taiwan," chaired by Joseph Gilmy.	George Washington University

In late 1989, it was announced that the Section would give an annual Outstanding International Accounting Educator Award beginning in 1990. Gary Meek chaired the first selection committee, which also was charged with developing guidelines for selection. Paul Garner was selected as the first recipient of the Outstanding International Educator Award (Garner is also a life member of the Section). Garner was a prime mover behind the requirement for curriculum internationalization by the American Academy of Collegiate Schools of Business. He has attended conferences and spoken in more than 80 countries.

Section Officers

The International Accounting Section has been blessed with a retinue of dedicated officers over its history. A list of those officers is on the following page. It should be noted that there was initially no provision for a practitioner vice president. Such a position was added when the bylaws were changed in 1978. Similarly, there was no separate position of treasurer until the change in bylaws. During the first two years of the Section's existence, the position was labeled secretary-treasurer. A six-member nominating committee is responsible for presenting a slate of prospective officers at each annual meeting. The nominations committee consists of the two most recent past Section chairpersons and four elected members.

Normally, members were given only one slate of candidates to vote upon at the annual meeting. However, at the 1987 meeting there were two nominees for academic vice chairman and secretary. As in prior years, the nominating committee for 1987 proposed one slate of candidates, which was headed by Kenneth Most of Florida International University. This slate was challenged by a minority of the nominating committee members who nominated a second slate. The second slate was headed

AAA INTERNATIONAL ACCOUNTING SECTION OFFICERS

YEAR	CHAIRMAN	1ST VICE CHR.	2ND VICE CHR	SECRETARY	TREASURER
1977	Hanns Martin Schoenfeld	Anita I. Tyra		Irving L. Fantl	
1978	Anita I. Tyra	Lee H. Radebaugh		Irving L. Fantl	
1979	Lee H. Radebaugh	Dhia D. Al Hashim	Ralph J. Mandarino	Irving L. Fantl	Norlin G. Rueschhoff
1980	Dhia D. Al Hashim	Norlin G. Rueschhoff	Alister K. Mason	John Brennan	Jane O. Burns
1981	Norlin Rueschhoff	Thomas Evans	Richard D. Fitzgerald	Konrad Kubin	Jane O. Burns
1982	Thomas Evans	Jane O. Burns	William Hayworth II	Konrad Kubin	Robert Raymond
1983	Jane O. Burns	Konrad Kubin	Robert Kleckner	Helen M. Gernon	Robert Raymond
1984	Konrad Kubin	H. Peter Holzer	Robert J. Sack	Helen M. Gernon	Abdel M. Agami
1985	H. Peter Holzer	Helen Gernon	Gary S. Schieneman	Belverd E. Needles, Jr.	Abdel M. Agami
1986	Helen Gernon	Abdel Agami	Walter F. O'Connor	Belverd E. Needles, Jr.	Adolf J. Enthoven
1987	Abdel M. Agami	Belverd E. Needles, Jr.	Roy C. Nash	Kenneth Most	Adolf J. Enthoven
1988	Belverd E. Needles, Jr.	Kenneth Most	Arthur Wyatt	Gary Meek	Adolf J. Enthoven
1989	Kenneth Most	Kathleen Bindon	Arthur Wyatt	Gary Meek	Juan Rivera
1990	Kathleen Bindon	Gary Meek	Alister Mason	Kathleen Sinning	Juan Rivera
1991	Gary Meek	Juan Rivera	Alister Mason	Kathleen Sinning	Maureen Berry

by Sidney Gray of the University of Glasgow. The slate headed by Kenneth Most won by a 27 to 23 vote. Most became the Section chairman for 1988-89.

Many committees have been quite productive over the years and have made major contributions to the field of international accounting. For instance, a 1977-78 committee sponsored a syllabus exchange program and made efforts to get international accounting course syllabi into Tom Burns' *Accounting Trends,* published by McGraw-Hill. Other committees, beginning in 1979, have sponsored pre-convention continuing education programs at the AAA annual and regional meetings.   The success of these pre-convention programs has varied over the years. For example, in the spring of 1984 there were workshops held at six of the seven regional meetings; only in the Southwest Region was there insufficient interest. In 1985, however, no workshops were held due to the low number of registrants ["Minutes of...," June, 1985, p. 2].

The first AAA/Price Waterhouse Doctoral Research Forum in International Accounting was held in Biloxi on the day preceding the 1984 Southeastern Regional meeting. Kenneth Austin, then at the University of Alabama, organized and coordinated the forum, which was expanded to other regions in subsequent years.

The International Accounting Section is now well established with a strong membership from throughout the world. The 1989 mission statement is indicative of the consideration being given to the future operations of the Section. The mission statement was accompanied by a list of 25 suggested actions which were brought forward by a Long Range Planning Task Force. If future officers will keep these suggested actions in front of them as a guide, the Section is destined for even greater successes in the future.

Conclusion

The American Accounting Association began its third quarter century with a committee acknowledgement that international accounting should be a part of the curriculum. This recognition mushroomed into a variety of programs designed to stress accounting as it is practiced outside of the United States. Nearly all aspects of the Association's activities now have an international orientation. Even the annual doctoral consortium now has foreign delegates. The International Lecturer program has been one of the AAA's most successful international outreach programs, despite its early difficulties. The main criticisms of the program have come from the international lecturers themselves who, to this day, feel that they are overscheduled. However, given the level of costs incurred for the program, it is probably one of the most successful AAA programs per dollar expended. To date, lecturers from the USA have visited no less than 45 foreign countries as a part of the program. Every inhabited area of the world has been visited.

Other international activities have been equally successful, but all of this success has not taken place without objections. Many members, including some of the past presidents, have complained that the Association's international programs have been enacted without any consideration of costs. Because the dues for international members have traditionally been less than for North American members, there has been some argument that members from the United States are supporting the foreign members. It has been pointed out that the lower dues (supposedly enacted because foreign members are less able to participate in Association activities) are insupportable given the higher costs of servicing the foreign members. For instance, several past presidents have pointed out that the postage for foreign mailings is greater than for domestic mailings, despite the fewer number of foreign members. Similarly, there has been criticism of the Executive Committee meeting in non-USA locations. Supporters of the AAA's international outreach programs argue that the current international environment requires the Association to operate in the international arena—an arena that hardly existed a quarter of a century ago.

Four past presidents, S. Paul Garner (president in 1951), David Solomons (1977-78), Stephen Zeff (1985-86), and Gerhard Mueller (1988-89) have been instrumental in the internationalization movement of the AAA. All but Garner have served as distinguished international lecturers, and he was the father of that program and has visited over 80 countries on his own initiative. These individuals made their international contributions not only while serving as president, but throughout the past quarter century. Today, the Association's international outreach programs are well established and can continue to grow under the direction of the officers of the International Accounting Section. Whereas it was individual efforts that provided the foundation for the AAA's international base, it seemingly will be group efforts that carry on and develop those efforts toward the Association's century mark.

References

"A Delicious Big Apple," *Accounting Education News*, October, 1986, pp. 5-7.

"AAA—International Accounting Section Mission Statement," *Forum*, Fall, 1989, p. 8.

Agami, Abdel M. and Felix P. Kollaritsch, *Annotated International Accounting Bibliography, 1972-1981*. Sarasota: American Accounting Association, 1983.

American Accounting Association Executive Committee, "Minutes of the Executive Committee Meetings, 1966-1989."

"Annual Meeting Program," *Accounting Education News*, June, 1986, pp. 5-15.

Bindon, Kathleen R., Letter to Gerald Searfoss, September 26, 1989.

Burns, Jane O., Michael A. Diamond, Helen Morsicato Gernon, *The International Accounting and Tax Researchers' Publication Guide*. Sarasota: American Accounting Association, 1982.

Byrd, Kenneth F., "Report for 1968-69," *Proceedings*, Canadian Regional Group, American Accounting Association, York University, June 7, 1969.

Chambers, Ray J., Letter to Paul Gerhardt, December, 1976.

Committee on International Accounting," International Dimensions of Accounting in the Curriculum," American Accounting Association, 1966.

Committee on International Accounting Operations and Education, *Accounting Education and the Third World*, Sarasota: American Accounting Association, 1978.

Committee on International Professor Exchange Program, "Committee Report," November 22, 1974.

Cushing, Barry E., ed., *Accounting and Culture* (Sarasota: American Accounting Association, 1986).

Education Committee of the International Accounting Section, "The Internationalization of Accounting Curriculum," American Accounting Association, 1978.

Flamson, Jack W., "International Auditing: Managing the Differences," *The Auditor's Report*, Fall, 1989, pp. 2ff.

Garner, S. Paul, "History of the International Section," *Forum*, January, 1985, p. 4.

Garner, S. Paul, Letter to Dale L. Flesher, January 5, 1990.

Garner, S. Paul, Letter to Dale L. Flesher, May 10, 1990.

Gerhardt, Paul, Letter to Kenneth Most, June 3, 1976.

Gerhardt, Paul, Letter to S. Paul Garner, November 12, 1976.

Gerhardt, Paul, Letter to Stephen A. Zeff, December 3, 1985.

Gibbins, Michael, Letter to Harold Q. Langenderfer, February 17, 1984.

Gibson, Robert W., "A Note of Recognition to the Initial Editor of *Forum*," *Forum*, January, 1986, p. 5.

"Hawaiian Annual Meeting an International Success," *Accounting Education News*, October, 1989, p. 9ff.

Hicks, Barry, Letter to Dale L. Flesher, June 26, 1990.

Horwitz, Bertrand, *Accounting Controls and the Soviet Economic Reforms of 1966*, Studies in Accounting Research No. 4, Sarasota: American Accounting Association, 1970.

Kubin, Konrad, "Chairperson's Message," *Forum*, November, 1983, p. 1.

Kubin, Konrad, "Chairperson's Message," *Forum*, March, 1984, p.2.

Miller, Herbert E., Letter to Paul Gerhardt, May 21, 1981.

"Minutes of Executive Committee Meeting," *Forum*, October, 1983, p. 4.

"Minutes of Executive Committee Meeting," *Forum*, June, 1985, p.6.

"Minutes of Executive Committee Meeting," *Forum*, October, 1985, p. 2.

Minutes of the Meetings of the Executive Committee, (Sarasota: American Accounting Association, 1966-1989).

Mueller, Gerhard G., "Report from the President," *Accounting Education News*, January, 1989, p. 3.

Needles, Belverd E., Jr., "Chairperson's Message," *Forum*, Winter, 1988, p. 1.

Needles, Belverd E., Jr., "Chairperson's Message, *Forum*, Summer, 1988, p. 1-2.

Needles, Belverd E., Jr., ed., *Comparative International Auditing Standards*. Sarasota: American Accounting Association, 1985.

Needles, Belverd E., Jr., Letter to David B. Pearson, August 20, 1987.

1989-90 Directory of the American Accounting Association. Sarasota: American Accounting Association, 1989.

Notable Contributions to the Periodical International Accounting Literature—1975-1978. Sarasota: American Accounting Association, 1979.

Previts, Gary John and Bruce Committe, *An Index to the Accounting Review, 1926-1978.* Sarasota: American Accounting Association, 1980.

"Report of the Committee on Accounting in Developing Countries," *The Accounting Review,* Supplement to Vol. LI, 1976, pp. 198-212.

"Report of the Committee on International Accounting," *The Accounting Review,* Supplement to Vol. XLVIII, 1973, pp. 121167.

"Report of the Committee on International Accounting," *The Accounting Review,* Supplement to Vol. XLIX, 1974, pp. 251-269.

"Report of the Committee on International Accounting," *The Accounting Review,* Supplement to Vol. L, 1975, pp. 91-95.

"Report of the American Accounting Association Committee on International Accounting," *The Accounting Review,* Supplement to Vol. LI, 1976, pp. 70-196.

"Report of the American Accounting Association Committee on International Accounting Operations and Education, 1975-1976," *The Accounting Review,* Supplement to Vol. LII, 1977, pp. 65-132.

"Report to the International Accounting Standards Committee from the Subcommittee on Depreciation Accounting," *The Accounting Review,* Supplement to Vol. LII, 1977, pp. 187-188.

"Report to the International Accounting Standards Committee from the Subcommittee on the Accounting Treatment of Changing Prices," *The Accounting Review,* Supplement to Vol. LII, 1977, pp. 191-196.

"Response to Exposure Draft Number 7 of the International Accounting Standards Committee Entitled, 'Statement of Source and Application of Funds'," *The Accounting Review,* Supplement to Vol. LII, 1977, pp. 199-204.

Rivera, Juan M. and Konrad W. Kubin, eds., *Proceedings of the V International Accounting Congress on Accounting Education.* Sarasota: American Accounting Association, 1987.

Sato, Seiichi, Kyosuke Sakate, Gerhard Mueller, Lee H. Radebaugh, eds., *A Compendium of Research on Information and Accounting for Managerial Decision and Control in Japan.* Sarasota: American Accounting Association, 1982.

Schoenfeld, Hanns Martin W., "The International Accounting Section of the AAA," *International Dimensions,* April, 1977, p. 1.

Schreuder, Hein, Letter to Paul Gerhardt, November 20, 1985.

Schweikart, James A., ed., *Cases in International Accounting.* Sarasota: American Accounting Association, 1987.

Scott, George M. and Pontus Troberg, *Eighty-eight International Accounting Problems in Rank Order of Importance—A DELPHI Evaluation.* Sarasota: American Accounting Association, 1980.

Sinning, Kathleen E., Letter to Steven M. Mintz, November 11, 1981.

Sinning, Kathleen E., ed., *Comparative International Taxation*, Sarasota: American Accounting Association, 1986.

Sommerfeld, Ray M., Letter to Belverd E. Needles, September 2, 1987.

Solomons, David, "The AAA-KMG International Conference on Standard Setting for Financial Reporting," *Accounting Education News*, November, 1986, pp. 5-7.

Stamp, Edward, Letter To Whom It May Concern, January, 1978.

Sterling, Robert R., Letter to Paul Gerhardt, July 14, 1976.

Tyra, Anita, Letter to Paul Gerhardt, March 13, 1978.

Zeff, Stephen A., Interview, December 11, 1989.

Zeff, Stephen A., Letter to Paul Gerhardt, July 27, 1977.

Zeff, Stephen A., "President's Message," *Accounting Education News*, October, 1985, pp. 1ff.

Zeff, Stephen A., "President's Message," *Accounting Education News*, November, 1985, pp. 4-5.

Zeff, Stephen A., "President's Message," *Accounting Education News*, May, 1986, pp. 7, 11.

CHAPTER 3
ANNUAL MEETINGS

For many members, the annual meeting is the primary activity of the American Accounting Association. Although this has always been true to some extent, it has become even more true in recent years as the readability level of the Association's primary journal, *The Accounting Review*, has changed, and as more premeeting seminars have been offered. Historically, when the annual meetings were held in December, the attendees were primarily the members themselves, without any accompanying family. However, when the meetings were switched to August, the time frame when families normally take vacations, the annual meetings became a family outing. Thus, today the attendance at the annual meetings by spouses and children typically exceeds that of members.

The 1966 annual meeting was the 50th anniversary meeting of the Association and was held at the Deauville Hotel in Miami Beach, Florida. Attendance was at an all-time record. Everett Royer of the University of Miami was in charge of local arrangements, and according to Herbert Miller, was responsible for much of the program planning. President Miller did arrange for the plenary and luncheon speakers

Past Presidents attending the 1966 50th Anniversary meeting included: (front row, l to r) Herbert E. Miller, Robert K. Mautz, Glenn A. Welsch, Walter G. Kell, Raymond C. Dein, A. B. Carson and Charles J. Gae; (center row, l to r) Herbert F. Taggert, Eric L. Kohler, G. H. Newlove, Russell A. Stevenson, William A. Paton and Fayette H. Elwell; (top row, l to r) Martin L. Black, Fr., C. Rollin Niswonger, C. A. Moyer, John A. White, Willard J. Graham, Frank P. Smith, S. Paul Garner, Harvey G. Meyer, and A. C. Littleton.

as is still customary to this day. One of the luncheon speakers was Lynn Townsend, president of Chrysler Corporation, who impressed the Executive Committee by arriving in a private plane, accompanied by a gun-carrying body guard. One of the plenary speakers was Robert Trueblood of Touche, Ross, Bailey & Smart who spoke on "The Role of Accounting in the Next Half Century." Raymond J. Chambers (University of Sydney) was also a plenary speaker with an address entitled "Prospective Adventures in Accounting Ideas."

With respect to convention by-products, they existed, but not to the extent they do today. There was a placement service, albeit small by today's standards. The exhibit area was not as heavily populated as it has been in recent years as only 13 publishers had booths (compared to 40 or more in recent years). Everett Royer also arranged for a post-conference cruise to the Caribbean which was quite popular. In fact, the trip's popularity encouraged Royer to arrange similar trips in future years. Of course Royer received a free trip for every 14 people who signed up, so he and his wife could go for free if 28 people signed up. People like Paul Garner and Charles Zlatkovich made these post-conference trips a regular part of their summer routine and as a result made many friends throughout the Association.

Note that the previous paragraph stated that the 1966 annual meeting was the 50th anniversary meeting; it was not, however, the 50th meeting because there were no national meetings held during 1942, 1943, and 1945 due to war-time restrictions and congested hotels. Regardless, the 1967 meeting was called the 51st annual meeting. This numbering system continued through the 1976 annual meeting which was billed as the 60th annual meeting. The 1977 meeting was never given a number in any published material other than the program booklet, which labeled the meeting as the 62nd annual meeting. The 1978 meeting was billed as the 63rd annual meeting. In other words, there was never a meeting labeled as the 61st annual meeting. Perhaps this new numbering system was adopted intentionally since the very first annual meeting had been held at the initial organizational meeting in December, 1916. Given that the first meeting was in 1916, the second in 1917, the third in 1918, etc., then 1977 would have been the 62nd annual meeting. In fact, the 1940 meeting was correctly billed as the 25th annual meeting and a big celebration was held. Because there were no meetings during 1942, 1943, and 1945 because of World War II travel restrictions, both the 1976 and 1977 numbering systems are incorrect. Actually, the 1966 annual meeting was the 48th meeting held. Thus, the 1976 meeting was the 58th meeting and the 1978 meeting was the 60th meeting held. Consequently, the meeting held in 1991 during the Association's 75th anniversary year will be the 73rd annual meeting (75 years + 1916 - 1942 - 1943 - 1945). Given that the program book for the 1990 meeting in Toronto proclaimed that meeting to be the 75th, it will be interesting to see what the program book for 1991 will call the Nashville meeting.

Annual Meeting Policies

The Executive Committee meeting held just prior to the 1966 meeting was witness to several new policy issues concerning annual meetings. The Long-Range Planning Committee had made ten recommendations, and these were passed unanimously. Included were the following provisions:

1. The annual meeting should be self supporting except for the cost of printing and mailing the preliminary announcement.
2. Ladies and children's programs should be provided but these and other family activities should not be subsidized ("ladies" programs originated at the 1948 meeting in Memphis and have been offered every year since).
3. The annual meeting should be held on a college or university campus during either the week preceding Labor Day or the second week before Labor Day.
4. Future convention sites should be selected five years in advance.
5. A "selected" future host school should be revisited eighteen months prior to the meeting by an inspection team appointed by the President.
6. Registration should begin at noon on Sunday.

Members of the 1966 Executive Committee: (l to r) Robert K. Jaedicke, Gordon Shillinglaw, Clarence L. Dunn, James B. Bower, Herbert E. Miller, Frank S. Kaulback, Jr., Lawrence L. Vance, Joe Fritzemeyer, Robert K. Mautz, Wendell Trumbull, and Harold Nelson.

7. The technical sessions should be held on Tuesday and Wednesday.

8. The president should have the responsibility for preparing the technical program.

9. The present practices in regard to publishers' exhibits should be continued.

10. Existing procedures in regard to unofficial social functions, i.e., open houses, should be continued [Minutes, August 20-21, 1966].

It should be noted that the term "ladies program" in item 2 above was a direct quote of the Long-Range Planning and Executive Committees, and not the words of this author.

Central Planning

In late 1977, Paul Gerhardt became concerned that the annual meetings had become too large to be organized by amateurs. The local arrangements committees had done excellent jobs over the years, but just when they learned how to run a convention, their job was over. Since there was no carryover from one year's committee to the next, there was no experience gained from the learning curve. After conferring with two previous local arrangements chairmen, Gerhardt recommended to the Executive Committee that it appoint a convention coordinator to handle conventions beginning with the 1979 convention in Hawaii. This would eliminate the need for a local arrangements committee. Gerhardt specifically recommend that the Adelle Cox Convention Services & Consultants of Miami be appointed to handle future conventions. The Executive Committee agreed, and Adelle Cox has worked pretty much full time for AAA since that time [Minutes, March, 1978, p. 19].

At about the same time, Gerhardt proposed that the timing and sites for future conventions be made more flexible. Instead of having the Executive Committee select the site of a future convention, Gerhardt suggested that the Committee select only the area of the country in which a future convention should be held and he would then investigate all possible facilities in an effort to obtain the best convention site and hotel. He felt that the Association's annual meetings had reached a magnitude which would justify competitive bidding among various cities and hotels for the Association's business. He also asked that he be given a two- or three-week window for the timing of future conventions in order to provide a better position for negotiating room rates. This proposal was also accepted by the Executive Committee [Minutes, August, 1978, p. 13].

Highlights of Annual Meetings

1967 Meeting—Pennsylvania State University, University Park, Pennsylvania—Lawrence Vance (University of California at Berkeley), President. President Vance was assisted in planning the program by vice president Gordon Shillinglaw.

A Professional Development Program on the use of math in accounting was held during the two weeks preceding the 1967 meeting. Plenary speakers included David Solomons (University of Pennsylvania), Walter G. Kell (University of Michigan), Hector Anton (Berkeley), and Peter A. Firmin (Tulane). Robert N. Anthony, then assistant secretary of defense, was a luncheon speaker.

1968 Meeting—San Diego State College Campus—Frank S. Kaulback (University of Virginia), President. Kaulback stated that he was solely responsible for the program. He did remember, however, that R. Lee Brummet recommended to him that a committee should be appointed to arrange the program for the annual meeting. Interestingly, six years later, it was Brummet who was the first president to actually appoint a technical program committee [Kaulback, October, 1989]. Vice president Peter Firmin had been assigned to assist in the preparation of the program [Minutes, August, 1967, p. 3]. The speakers at the opening plenary session were Leonard P. Spacek of Arthur Andersen & Co. and Marvin L. Stone, then president of the AICPA. Comptroller General Elmer B. Staats was a luncheon speaker. The remainder of the two-day program was composed of twelve concurrent sessions and a panel discussion.

1969 Meeting—University of Notre Dame, South Bend, Indiana—Sidney Davidson (University of Chicago), President. Vice president Tom Keller (Duke University) was to assist with the program. As had been true of previous meetings, much of the content consisted of panel discussions of AAA committee reports. Early arrivals at the banquet on the final night found the door to the banquet facility locked. After some delay, Ray Powell (University of Notre Dame) came back with the key, and all was well. Future AAA president R. Lee Brummet headed a plenary session with his report on the subject he was to become famous for, "Accounting for Human Resources." Another plenary session was shared by Robert M. Trueblood (Touche, Ross, Bailey & Smart) and John T. Wheeler (Berkeley). AICPA president Ralph E. Kent (Arthur Young & Co) was a luncheon speaker. For the first time in this quarter century, there was a concurrent panel session on accounting history. The panelists who discussed "A Role for Accounting History" were Stephen A. Zeff (Tulane), David F. Hawkins (Harvard), and Maurice Moonitz (Berkeley).

1970 Meeting—University of Maryland, University Park, Maryland—Norton Bedford (University of Illinois), President. The opening reception was held on a boat. AICPA president Louis M. Kessler (Alexander Grant & Co.) was a luncheon speaker. A plenary session on accounting education was shared by George Sorter (University of Chicago), R. Lee Brummet (North Carolina), and Charles Horngren (Stanford). Another plenary session, shared by six speakers, dealt with current trends in accounting education and research. Included among the six speakers were Gordon Davis (University of Minnesota) who spoke on the "Role of Computers in Accounting," and David Green (University of Chicago) whose subject was "Empirical Research Developments." In many respects the subject matter of this conference seemed to indicate a turning point in the direction of accounting education and research. There was more emphasis on quantitative methods, foundations of accounting measurement, socioeconomic accounting, behavioral science, and emerging trends. In substance, the subject matter of the 1970 meeting seemed more akin to the subject matter of the present day than did that of earlier conventions. Also, there were fewer practitioner speakers than had been true in prior years.

1971 Meeting—University of Kentucky, Lexington, Kentucky—James Don Edwards (Michigan State University), President. President Edwards issued a call for papers for the annual meeting—the first time that such a procedure was used. The call for papers was sent to about 800 members. Edwards stated that he accepted every paper submitted except for one. That one paper was submitted by someone who had been on the program the two preceding years and Don was worrying that the individual was getting too much exposure. He later said that he was sorry that he had not accepted that paper [Edwards, October, 1989]. Perhaps as a result of the call for papers, the 1971 meeting had many more sessions and more participants than in previous years. The meeting began on Monday afternoon instead of the traditional Tuesday morning. There were five concurrent sessions, labeled "Special Interest Groups," on Monday afternoon, followed by a buffet dinner at nearby Keeneland Race Course. Whereas there had been 23 authors and 24 panelists who participated in the 1970 meeting, the 1971 meeting featured 56 authors and 56 panelists. Speakers at the first plenary session were Richard T. Baker (Ernst & Ernst), Terrance Hanold (president, Pillsbury Company), and David Solomons (University of Pennsylvania). The Wednesday plenary session was headed by Secretary of Commerce Maurice H. Stans (this was the year before his involvement in the Watergate situation). The luncheon speaker was AICPA president Marshall Armstrong (George S. Olive & Co). The speaker at the closing banquet was the legendary Kentucky basketball coach Adolph Rupp. In size and substance, the 1971 meeting was beginning to look like the meetings of the 1980s.

1972 Meeting—University of Utah, Salt Lake City, Utah—Charles T. Zlatkovich (University of Texas), President. Zlatkovich said that preparing for the meeting was the biggest part of his job—he had to visit Salt Lake City six times during the year prior to the meeting in order to make arrangements [Zlatkovich, March, 1990]. Again the meeting began with meetings of special interest groups on Monday afternoon. One of the sessions was entitled a "Rap Session With the Executive Committee," the purpose of which was to give members an opportunity to express their viewpoints regarding the activities of the Association. The first plenary session featured Francis M. Wheat discussing the work of the Accounting Principles Study Group (the Wheat Committee), Frank T. Weston (Arthur Young & Co.), and Abraham J. Briloff (City University of New York). Briloff's speech carried the most interesting title of any at the convention—"Between Scylla and Charybdis."

1973 Meeting—August 15-17, Universite Laval, Quebec City, Quebec—Robert T. Sprouse (Stanford University), President. Fernand Sylvain (Universite Laval) was chairman of the convention arrangements committee. Records were broken as 1,429 members attended. This was one meeting where the highlight of the week was the business meeting. There was a spirited discussion at the business session as to whether the AAA should have "official" positions on accounting principles. The Executive Committee had essentially been unanimous, with the exception of Robert Mautz, that the Association should indeed issue official positions. The Executive Committee thought that the issue would be approved by the membership without controversy. However, Mautz stood up at the business meeting and convinced those in attendance that the AAA should not speak with one voice. The members defeated the proposal, much to the surprise of the Executive Committee. An issue that was passed was to increase annual dues to $25. The Association had closed its 1973 fiscal year with a deficit of $45,965, which was over 25 percent of annual membership dues. Plenary session speakers included three Canadians: R. C. Scrivener (Bell Canada), Robert M. Fowler (Private Planning Association of Canada), and Frank Capon, past president of the Canadian Institute of Chartered Accountants ["Annual Meeting A Great Success," 1973, p. 1].

1974 Meeting—August 19-21, Fairmont Hotel, New Orleans, Louisiana—Robert N. Anthony (Harvard University), President. For the first time since 1966, the meeting was held in a hotel instead of on a campus. The convention arrangements committee, chaired by Stephen A. Zeff (Tulane University), consisted of the faculty and wives from Tulane University, Southern University in New Orleans, and Louisiana State University in New Orleans (now University of New Orleans). Planning for this meeting was complicated somewhat by the gasoline shortage that existed in the winter of 1973-74. The arrangements committee could not decide how much the registration fee should be because they had no idea whether anyone would be able to get gas to come to the meeting. Since many of the hotel costs were fixed costs, a small attendance would have meant a major loss for the Association. There was even fear that the annual meeting would have to be cancelled because of travel restrictions. However, all of the worrying was for naught as gasoline supplies stabilized and the conference was held without any problems [Zeff, December, 1989]. Over 1,450 members attended. A Professorial Development Course on Behavioral Science, led by James E. Sorensen of the University of Denver, was held the week preceding the meeting in Baton Rouge. There were also two two-day courses offered prior to the meeting. One of these dealt with research appreciation and was taught by Edward L. Summers (University of Texas). The other, dealing with accounting principles and auditing concepts was taught by Doug Carmichael (AICPA) and Arthur Wyatt (Arthur Andersen & Co.). There were two plenary sessions and 23 concurrent sessions. Plenary and luncheon speakers included C. Jackson Grayson, Jr. (Southern Methodist), Robert K. Mautz (Ernst & Ernst), Howard W. Wright (University of Maryland), Marshall S. Armstrong (chairman, FASB), John C. Burton (SEC), Sidney Davidson (University of Chicago), and Samuel A. Derieux (president, AICPA). One of the most talked about, and perhaps most enduring, aspects of the meeting was the announcement by Gerald Polanski of the Touche Ross Foundation that the Trueblood Seminars had been established and the Foundation was pledging a quarter of a million dollars for the first three years of seminars. Social activities were highlighted by a Mardi Gras ball on Monday evening ["1974 Annual...," 1974, p. 2].

1975 Meeting—August 18-20, Tucson Community Center, Tucson, Arizona— R. Lee Brummet (University of North Carolina), President. Hosted by the University of Arizona with William B. Barrett as chairman of the arrangements committee. Lee Brummet established a Technical Program Committee, which he chaired, to help with the planning of the program. This was the AAA's first use of a program committee for the annual meeting. Over 1,400 members registered for the meeting. Although the Braniff Place Hotel was adjacent, most members had to be bussed a long way from their motels. A tennis resort was quite popular among those who brought their families. In addition to having both plenary and concurrent sessions, there was the introduction of a "research carnival" where each participant was assigned a carrel with space to accommodate a limited number of ad hoc discussants. Those attending the carnival could move about from carrel to carrel as their interests dictated ["President's Call...," 1975, p. 1]. This concept was reintroduced by Gary Previts at the 1990 meeting in Toronto. One notable introduction at this meeting was the publication of proceedings. Paul Gerhardt initiated the idea and was authorized at the November 1974 Executive Committee meeting to publish a volume of proceedings with a copy being distributed to each member attending the annual meeting. The publication of the annual meeting proceedings became a tradition in following years. The practice of offering continuing education programs on the two days prior to the meeting was enhanced significantly in 1975 as seven different programs were offered. There were two plenary sessions and 37 concurrent sessions ["Tucson...," 1975, p. 2]. Plenary session speakers included Alva O. Way (General Electric), Sidney Jones (Counselor to the Secretary of the Treasury), Frank Zarb (Federal Energy Administrator), Herbert E. Miller (Arthur Andersen), Peter Firmin (University of Denver), and Justin Davidson (Cornell University) ["Advance...," 1975, p. 6-7]. The main social event was a dinner at Old Tucson, a realistic western town often used by movie production companies.

1976 Meeting—August 23-25, Hyatt Regency, Atlanta, Georgia, Georgia State University was the sponsor—Wilton T. Anderson (Oklahoma State University), President. Catherine Miles (Georgia State University) chaired the local arrangements committee. More people than usual came early for this meeting because the Second World Congress of Accounting Historians was held the day preceding the beginning of the annual meeting and the Association sponsored eight pre-meeting continuing education courses. More memorable than the World Congress or the continuing education courses, however, was the blood bath that several members received the night before the conference began when a very plump woman (who was not affiliated with AAA) jumped off an upper balcony of the Hyatt Regency and plummeted to her death in the lobby below. Several AAA members were splattered with blood when the woman landed. Needless to say, it was a traumatic experience for those relaxing in the lobby. This was the year of the U. S. bicentennial and it was quite a coincidence when the number of members who registered for the meeting totaled a record 1,776. Some people claimed that the Georgia State faculty closed the registration booth when the number reached the proper historic total. The first plenary session centered around international accounting and featured speakers Adolf Enthoven (University of Texas at Dallas), Guy Barbier (Arthur Andersen), and Louis M. Kessler (Alexander Grant & Co.). Other plenary speakers included Richard E. Flaherty (University of Illinois), Norton M. Bedford (University of Illinois), and Albert H. Cohen (Price Waterhouse) ["Come...," 1976, p. 3]. Members were also treated to hear two other accountancy legends in the persons of Raymond J. Chambers (University of Sydney), the AAA Distinguished International Lecturer, and William A. Paton, then 87 years old, one of the founders of the Association, who had not attended an annual meeting in seven years. Paton, who acknowledged that he still hated Franklin D. Roosevelt, received a standing ovation from the Tuesday luncheon crowd ["Mr. Accounting," 1976, p. 5]. Another activity that took place for the first time was the meetings of the special interest sections. Six sections held organizational meetings on Monday afternoon. Social activities included a choice of two dinner theaters and an indoor entertainment park called the World of Sid and Marty Krofft. The closing banquet was preceded by a reception sponsored by the publishing firm of Richard D. Irwin. This was the first year that Irwin sponsored the reception wherein all registrants were invited. In earlier years, the Irwin reception had been more selective and was held away from the convention functions ["Record...," 1976, p. 1-2].

1977 Meeting—August 22-24, Hilton Hotel, Portland, Oregon—Charles T. Horngren (Stanford University), President; James Fremgen (Naval Postgraduate School) was chairman of the technical program committee. Portland State University was the host school, but a local firm was appointed the official convention coordinator. James Bentley of Portland State University was chairman of the local arrangements committee. Bentley was a replacement for his colleague Joseph J. O'Rourke who was the initial chairman, but died prior to the convention ["New Notes," 1977, p. 9]. There were 1,602 members registering plus about 1,500 family members. Due to the small size of the Hilton and other hotels in the area, members were spread out over several hotels. Plenary sessions were held in the Fox Theater across the street from the Hilton. For the first time, Monday afternoon was devoted to meetings of the new special interest sections of the Association. The Monday evening buffet at nearby Washington Park featured Columbia River salmon cooked over an open fire; however, by the time most members were served, any food would have been welcome. The delays in serving the meal had members joking that they were worried about missing their Thursday flights. Entertainment featured lumberjacks and Indians. President Chuck Horngren and past president Sidney Davidson arrived at the cookout on the last bus from the hotel. Chuck claimed that the food line was the longest line he had ever seen and was moving at the rate of about four feet every twenty minutes. He and Sidney and their spouses returned to the hotel to eat—where the service was also slow. The entertainers at the closing banquet on Wednesday evening selected volunteers (unwilling volunteers) from the audience, and somehow were able to get past president Herbert E. Miller to perform a combination comedy and dance routine. He was great.

1978 Meeting—August 21-23, Hilton Hotel, Denver, Colorado—David Solomons, President; Gary John Previts (University of Alabama), chairman of Technical Program Committee. Ray L. Brown (University of Denver) was chairman of the local arrangements committee. This meeting was originally scheduled for Kansas City, but was changed in 1976 due to the rundown condition of the downtown area and the poor condition of the facilities where the convention was being planned [Minutes, March, 1976, p. 12]. Attendance in Denver was again at an all time high with 1,888 members attending, plus 153 exhibitors and about 2,000 family members ["Sixty-Third, 1978, p 1]. One of the highlights was that David Solomons had arranged for the plenary sessions to be held in the Centre Theater next door to the convention hotel. However, a few minutes before the session was to begin, he learned that the theater doors were locked because the theater wanted a prepayment for the use of the facilities. Fortunately, Solomons was able to assure the theater's management that they would get the money soon. However, he did take a partner from Arthur Young & Co. with him in case he needed money. The partner had agreed to write a check on the spot if that was what it took to get the doors open [Solomons, December, 1989]. Seven continuing education courses were offered on the days preceding the meeting. On the Sunday prior to the conference was the first meeting of the new Advisory Council, a group which was later destined to play a more significant role in the governance of the organization. Plenary sessions included one on "The Accounting Profession and Self Regulation" headed by John

Biegler, senior partner of Price Waterhouse & Co., and Harold Williams, chairman of the SEC. Another plenary session featured Oscar Gellein of the FASB and Victor H. Brown of Standard Oil of Indiana [Sixty-Third, 1978, p. 1].

1979 Meeting—Sheraton-Waikiki Hotel, Honolulu, Hawaii—Maurice Moonitz (University of California at Berkeley), President; Carl G. Orne (California State University at Hayward) was the chairman of the Technical Program Committee. Moonitz stated that he selected Orne, then a relatively unknown entity in AAA circles, to develop the program because he had worked with Orne in the California State Society of CPAs and knew that he was a take-charge type of person who could get the job done. Also, he lived fairly close to Moonitz and the two could get together for lunch periodically. Orne later told Moonitz that being program chairman was the best thing that ever happened to his career; the national publicity Orne received got him promoted to full professor [Moonitz, 1990]. Only 1,341 members registered for the meeting—a significant decrease from prior years due to the high costs of reaching the island state. Eight preconference continuing education programs were scheduled, but six had insufficient registrants and had to be cancelled. Plenary speakers included Nils H. Hakansson (California-Berkeley), Thomas Dyckman (Cornell), Elmer Staats (Comptroller General of the U.S.), William W. Cooper (Harvard), Chester Vanatta (Arthur Young & Company), and John C. Burton (Columbia University).

1980 Meeting—August 11-13, Sheraton Boston Hotel, Boston, Massachusetts—Donald H. Skadden (University of Michigan), President; Harold Q. Langenderfer (University of North Carolina) was chairman of the program committee. Over 1,800 members attended. Eleven continuing education courses were held prior to the annual meeting. There were 47 concurrent sessions and two plenary sessions. Plenary speakers were Jerome Kurtz (Commissioner of Internal Revenue), Washington SyCip (SGV Group, Manila), Donald J. Kirk (FASB), Charles Horngren (Stanford), and Reed K. Storey (FASB). At the banquet Administrative Secretary Paul L. Gerhardt received a special award in recognition of 15 years of service to the AAA. The annual banquet was held on Tuesday evening, rather than Wednesday, and the last sessions were completed by 3:30 p.m. on Wednesday to allow many of the attendees to return home that same day. The placement activities were organized by Robert J. Hehre (Northeastern University), who was to provide a similar service for several years to come ["Boston..., 1980, p. 6-7].

1981 Meeting—Hyatt Regency, Chicago, Illinois—Joseph A. Silvoso (University of Missouri), President; Doyle Z. Williams (University of Southern California) was program chairman. There were 52 concurrent sessions and two plenary sessions. Plenary speakers were William S. Kanaga (Chairman, AICPA), John Shad (SEC), Allan Cook (International Accounting Standards Committee), Michael

O. Alexander (FASB), and Andrew McCosh (Manchester Business School). Luncheon speakers were William C. Freund (New York Stock Exchange) and David Brinkley (NBC News). Some members had difficulty reaching the meeting because the nation's air traffic controllers went on strike a few days before the meeting began. Still, 1,807 members were in attendance. There were about 80 no-shows among those who preregistered. The welcome reception featured clowns, magicians, and jugglers. The Friday banquet entertainment included comedian Henny Youngman and the Serendipity Singers ["66th Annual...," 1981, p. 1].

1982 Meeting—Town and Country Hotel, San Diego, California—Thomas R. Dyckman, President; Gary Sundem (University of Washington) was program chairman. This year marked another record attendance. There were 30 alumni get-togethers, 58 concurrent sessions, and two plenary sessions. The plenary speakers were Michael C. Jensen (University of Rochester), Robert Wilson (Stanford), Stephen B. Ross (Yale), and Karl Weick (Cornell). Social activities included a visit to Sea World. Luncheon speakers included economist Arthur B. Laffer and newspaper columnist Jack Anderson. Entertainment at the closing banquet featured comedian Don Rice III and the New Christy Minstrels ["San Diego...," 1982, pp. 4-5].

1983 Meeting—Marriott Hotel, New Orleans, Louisiana—Yuji Ijiri (Carnegie Mellon University), President; Roland "Pete" Dukes (University of Washington) was program chairman. Registrants totaled 2,036, plus spouses and children. For most attendees, the memorable aspect of the New Orleans meeting was either the long lines that existed at check-in or the elevator service (or lack of it) at the Marriott. Few people were able to check in to their rooms without at least a two-hour wait. The Marriott had only two or three clerks to service a couple of thousand irritated AAA members. Once the meeting got started, however, the check-in fiasco gave way to the elevator fiasco. At times, there were 25-minute waits for an elevator. No one ever figured out where the six elevators went once they left the ground floor. At one time, however, Tonya Flesher (University of Mississippi) and Andy Barr (former SEC chief accountant) thought they were fortunate in quickly getting an elevator. Their fortune soon changed to misfortune when the elevator got stuck between floors. Hotel employees finally pried the doors open a half hour later. From then on, Tonya took the stairs to her seventh floor room. Most other members housed on the first ten floors took similar measures. The real highlight of the meeting was the "Outstanding Educators" videotape that was shown at the Tuesday luncheon. Also, Carl Devine and William A. Paton, Sr. were given Outstanding Educator Awards. Paton wowed the audience with his acceptance speech and received a standing ovation. Plenary speakers included Arno A. Penzias (Bell Laboratories and a Nobel Laureate in physics), Dana R. Richardson (Arthur Young & Company), Maurice Moonitz (California—Berkeley), Robert Kaplan (Carnegie-Mellon

University), and Steve Zeff (Rice University). Newsman Howard K. Smith spoke at the closing luncheon on Wednesday ["Sixty-Eighth...," 1983, pp 4-6].

1984 Meeting—Toronto, Ontario—Harold Q. Langenderfer (University of North Carolina), President; John K. Simmons (University of Florida) was the chairman of the program committee. As was true the previous time the Association met in Canada, annual dues were increased. Various computer courses were offered in the premeeting CPE program. Albert J. Arsenault, Jr. (Hillsborough Community College) had secured ninety computers for on-site use by CPE participants. The meeting featured two plenary sessions and 61 concurrent sessions. One of the plenary session speakers was the Rt. Hon. Edward R. G. Heath, former prime minister of Great Britain. The other plenary session featured AICPA president Philip C. Chenok. Newspaper columnist Georgie Anne Geyer was the luncheon speaker ["Toronto...," 1984, p. 6-9].

1985 Meeting—MGM Grand Hotel, Reno, Nevada—Doyle Z. Williams (University of Southern California), President; Jan R. Williams (University of Tennessee) was chairman of the program committee. A new record attendance of over 2,200 members showed up in Reno. This was over 20 percent of the Association's membership. Part of the attraction may have been because lodging costs were lower than had been the case in recent years since the hotel was also a casino. One interesting anecdote that occurred during the meeting involved one of the many honeymoon suites in the hotel. Since Reno is known for its wedding chapels, the hotel had an abundance of honeymoon suites with large round beds. When Arthur Mehl of Bradley University checked in late, he was given a honeymoon suite because the hotel was out of regular rooms. Professor Mehl got a lot of kidding because of his sexy round bed. He was not sure where the pillows were supposed to go. Unfortunately (or fortunately ?), Art's wife was back in Peoria. Plenary speakers included James B. Maas (Psychology professor at Cornell), Jim Antonio (GASB), Ray J. Groves (AICPA chairman), and John M. Fedders (SEC). Hugh Sidey of Time Magazine was the Wednesday luncheon speaker. Many in attendance felt that this was one of the Association's best annual meetings ever because of the quality of the program and the fine facilities. In fact, for those who did not gamble, the costs were quite low ["Reno...," 1985, pp. 4-5].

1986 Meeting—New York City, New York—Steve Zeff (Rice University), President; Robert E. Jensen (Trinity University) was chairman of the Technical Program Committee. Zeff planned the dates of the meeting so that members could stay over Saturday night and get cheaper airline fares. Instead of a closing banquet, Zeff arranged for members to attend the Broadway play, "Cats" at the Winter Garden Theater. This was a popular diversion for over 1,300 who attended. Lynn Seidler, wife of AAA member Lee Seidler and executive director of the Schubert

Foundation, assisted the Association in buying out the theater. Attendance at the meeting reached a record 2,427 members. There were three plenary sessions, the speakers for which were Geert Hofstede (Netherlands), Hein Schreuder (University of Limburg), James G. March (Stanford), Anthony G. Hopwood (London School of Economics), Robert S. Kaplan (Harvard and Carnegie-Mellon), and Joel Demski (Yale). In addition, there were 72 concurrent sessions. Julian Paleson (Grant Thornton Europe) was a luncheon speaker ["A Delicious...," 1986, p. 5].

1987 Meeting—Albert Sabin Convention Center (Clarion and Hyatt Regency Hotels), Cincinnati, Ohio—Ray Sommerfeld (University of Texas), President; John L. Kramer (University of Florida) was program chairman. The highlight of this event was the Tuesday evening baseball game at Cincinnati's Riverfront Stadium. Ray Sommerfeld and Bill Beaver threw out the first ball before the game between the Reds and the Pittsburgh Pirates. Two AAA representatives, Ed Currie (University of San Francisco) and Laurie Pastena (wife of member Victor Pastena, City University of New York), sang the national anthem before the game. Member attendance reached 2,306, with over 600 members coming early for the Sunday continuing education programs. The 2,306 members registered was the second highest ever as only the 1986 meeting had a greater attendance. The first plenary speaker was Donald J. Kirk (Columbia University), who discussed his years on the FASB. As is befitting a tax professor, President Sommerfeld arranged the second plenary session around the subject of taxation. The speakers were economist Joseph E. Stiglitz (Princeton University) and Mark A. Wolfson (Stanford). AICPA chairman J. Michael Cook (Deloitte Haskins & Sells) was the Tuesday luncheon speaker ["The 72nd...," 1987, pp. 4-9].

1988 Meeting—Orlando, Florida—William Beaver (Stanford University), President; Daniel W. Collins (University of Iowa), was chairman of the Technical Program Committee. All AAA annual meetings are family affairs, but this one was more so than ever. In fact, it seemed impossible to find the people you wanted to see because they were mostly out at Disney World instead of attending the conference sessions. The Monday evening welcome reception was enjoyed by over 3,500 members and their families. That reception had an Everglades theme including large live alligators, which could be petted (they had their mouths taped shut). Speakers at the plenary sessions included Thomas Dyckman (Cornell), Robert T. Sprouse (San Diego State University), Rick Antle (Yale), and George Foster (Stanford). John C. Burton (Columbia University) was a luncheon speaker. In a departure from previous practice, the Outstanding Educator Award winners, Robert S. Kaplan and Stephen Zeff, were asked to make more elaborate remarks than had been requested of award recipients in past years. Each gave 15-minute addresses to the luncheon attendees. Over 2,400 members attended the meeting, plus a record number of exhibitors. A silent auction held for the benefit of the Fellowship Fund raised over $10,000. Prizes were donated by exhibitors ["1988 Annual...," 1988, pp. 8-11].

1989 Meeting—Hilton Hawaiian Village, Honolulu, Hawaii—Gerhard Mueller (University of Washington), President; Douglas A. Snowball (University of Florida), chairman of Technical Program Committee. Registrants totalled 1,890. As would be expected with a meeting held in the international city of Honolulu and presided over by a famed international accountant, the 1989 meeting was international in scope. In fact, the members of the technical program committee had been instructed to have as many international sessions as possible. The most common subject of discussion was how the travel agents had mishandled members' airline and hotel reservations. Everyone had a different story to tell—all bad. However, once people arrived in Honolulu, there was nothing but praise for the meeting and the facilities. Again the annual dues were increased. Past president Steve Zeff thought the decision on raising dues should be postponed one year because the Association had a tradition of increasing dues every year that the meeting was held in Canada. However, the budget deficit took precedence over tradition. The preconference continuing education programs, 21 of them, were held at the Royal Waikoloan Hotel on the big island of Hawaii and attracted about 300 registrants. One of the highlights was the opening ceremony which featured representatives from eight AAA associate organizations. Plenary speakers included Charles E. Young (UCLA Chancellor), Norishige Hasegawa (Sumitomo Chemical Co, Japan), and Sidney J. Gray (University of Glasgow). Robert L. May (Arthur Andersen & Co.) was a luncheon speaker. The Farewell Banquet on Wednesday night featured entertainment by Hollywood personality (hoochey-koochey dancer) Charo. The silent auction was not as successful as in the previous year, but still raised nearly $5,000 for the Fellowship Fund ["Hawaiian...," 1989, pp. 9-15].

1990 Meeting—August 8-11, Sheraton Hotel, Toronto, Ontario—John Simmons (University of Florida), President; Gary John Previts (Case Western Reserve University), chairman of Technical Program Committee. Again, the members were impressed with the cleanliness of the Canadian city and the relatively crime-free downtown area. The major disappointment was that the Toronto Bluejays baseball team, playing their first season in a new domed stadium, were essentially sold out for all games. Member participation was at an all-time high with 2,588 members registering. There were three plenary sessions and 78 concurrent sessions. In addition, there were 52 papers scheduled for a "Research Forum," wherein authors sat around tables and discussed their research with whomever happened to come around. Counting discussants and moderators, there were a total of 625 participants' names in the program booklet. Plenary session speakers included Stephen Lewis, former Canadian Ambassador to the United Nations, J. Michael Cook (Deloitte & Touche), Christopher J. Steffen (Honeywell, Inc.), Doyle Williams (University of Southern California), Shyam Sunder (Carnegie Mellon University), and Robert Elliott (KPMG Peat Marwick). Economist Lester C. Thurow was a luncheon speaker, as was AICPA president Philip B. Chenok ["The 75th....," 1990].

1991 Meeting—August 11-14, Nashville, Tennessee—Alvin A. Arens (Michigan State University), President. Fred Neumann (University of Illinois) was program chairman. The highlight of this meeting was the distribution of the volume commemorating the 75th anniversary history of the Association.

Summary and Conclusion

Although the 1966 annual meeting was held in a large Miami hotel, subsequent meetings returned to campuses, which was an official policy for many years. Unfortunately, as the attendance grew, the campuses could no longer provide the facilities needed by the Association. Thus, the 1973 meeting in Quebec City was the last to be held on a university campus.

Historically, the planning of the annual meeting was the responsibility of the president, but beginning in 1975, a program committee has taken much of the burden off of the president. Of course it might be argued that the technical program would not have been such a burden had not James Don Edwards widened the level of participation at the 1971 meeting. The individuals who have served as chairmen of the program committee have for the most part been well known leaders of the profession. In fact, four of the first eight chairmen later became Association president. Others later joined the Executive Committee in other roles. Thus, the position of program chairman offers the opportunity to serve the Association in a very visible role.

A local arrangements committee did much to help the president in preparing for the annual meeting prior to 1978. However, the need for such a committee was sometimes viewed as a reason for schools not to offer to sponsor future meetings. As the meetings grew larger, it became necessary to replace the local arrangements committees with professional meeting planners. Thus, beginning in 1979, Adele Cox of Miami, Florida, has been responsible for planning the locations of AAA annual meetings (and many regional meetings).

The annual meeting has always been the main activity of the Association, and the size of that annual meeting has grown ever larger in recent years. Members have the opportunity to visit all parts of the country over a period of years, and to have such vacations either subsidized by their employer or made tax deductible. In addition to the regular meeting activities, there are also section meetings and preconference continuing education programs. Combine these with committee meetings, the need to meet with coauthors and graduate school colleagues, and an abundance of recruiting activities, and the annual meeting becomes a nightmare of scheduling for many members. At the same time that the member is trying to attend all sorts of professional activities, there are spouses and children clamoring for attention because they think of the meeting as the annual summer vacation.

Despite this abundance of meetings, members leave the annual conference refreshed and ready to face another school year. What is it that makes the meetings so exciting? There is something emotional about coming in contact with many of the legends of the profession—to be able to rub shoulder to shoulder with famed authors, noted theoreticians, and partners from large accounting firms. This author remembers the excitement of that first meeting with William A. Paton, John Carey, Charles Horngren, even Richard D. Irwin. I never realized that there actually was a person named Richard D. Irwin until he came up and introduced himself at the 1969 annual meeting. In summary, the annual meeting is the Association's major membership activity, and is loved by all, but thankfully comes only once a year.

References

"A Delicious Big Apple," *Accounting Education News*, October, 1986, pp.5-7.

"Advance Technical Program," *Accounting Education News*, April, 1975, pp. 6-7.

"Annual Meeting A Great Success," *Accounting Education News*, October, 1973, pp. 1-2.

"Boston Meeting a Great Success," *Accounting Education News*, November, 1980, pp. 6-7.

"Come to Atlanta," *Accounting Education News*, June, 1976, p. 3.

Edwards, James Don, Interview by Terry K. Sheldahl, October 1, 1989.

"Hawaiian Annual Meeting An International Success," *Accounting Education News*, October, 1989, pp. 9-15.

Horngren, Charles, Interview by Dale L. Flesher, July 28, 1990.

Kaulback, Frank, Interview by Terry K. Sheldahl, October 5, 1989.

Miller, Herbert E., Interview by Tery K. Sheldahl, October 1, 1989.

Moonitz, Maurice, Interview by Dale L. Flesher, July 27, 1990.

"Mr. Accounting," *Accounting Education News*, November, 1976, pp. 2 and 5.

"News Notes," *Accounting Education News*, January, 1977, p. 9.

"1988 Annual Meeting," *Accounting Education News*, October, 1988, pp. 8-11.

"1974 Annual Meeting Report," *Accounting Education News*, October, 1974, p. 2.

"1979 Annual Meeting," *Accounting Education News*, November, 1979, pp. 1-2.

"President's Call for Papers and Abstracts for Tucson Meetings," *Accounting Education News*, March, 1975, p. 1.

"Record Turnout in Atlanta," *Accounting Education News*, November, 1976, pp. 1-2.

"Reno Draws Record Number of Members," *Accounting Education News*, October, 1985, pp. 4-5.

"San Diego Convention Draws Record Attendance," *Accounting Education News*, October, 1982, pp. 4-5.

"Sixty-Eighth Annual Meeting Draws Record Attendance," *Accounting Education News*, October, 1983, pp. 4-6.

"66th Annual Meeting a Great Success," *Accounting Education News*, October, 1981, pp. 1-3.

"Sixty-Third Annual Meeting Draws Record Attendance," *Accounting Education News*, November, 1978, pp. 1-2.

Solomons, David, Interview by Terry K. Sheldahl, October, 1989.

"The 72nd Annual Meeting," *Accounting Education News*, October, 1987, pp. 4-9.

The 75th Annual Meeting of the American Accounting Association, Program for 1990 annual meeting in Toronto, 1990.

"Toronto Annual Meeting Attendance Tops 4,000," *Accounting Education News*, October, 1984, pp. 6-9.

"Tucson—Scene of Successful Annual Meeting," *Accounting Education News*, November, 1975, pp. 2-3.

Zeff, Stephen A., Interview by Dale L. Flesher, December, 1989.

Zlatkovich, Charles, Interview by Dale L. Flesher, March, 1990.

CHAPTER 4
THE AAA
REGIONAL GROUPS

Although the American Accounting Association (AAA) was founded in 1916, regional meetings were not held until 1949 when the Southeastern Region group held its first meeting at Georgia Tech University in Atlanta. Today, there are seven regional groups within the AAA structure as the entire country has been divided into regions. The success and influence of the Southeastern Region meetings was instrumental in this overall regionalization of the AAA. Regionalization, in turn, has led to a variety of changes in organizational structure within the Association.

In 1952, following the early success of the Southeastern Regional Group, the AAA Executive Committee established a subcommittee on regional meetings made up of Richard Claire of Arthur Andersen & Co. and S. Paul Garner of the University of Alabama. The consensus of that subcommittee was that since the membership of the organization had grown to over 4,500 people, there was a need to broaden geographical representation in order to enhance membership participation. The main recommendation was that the AAA should not "push" regional developments, but let them develop naturally and in a self supporting way. The subcommittee argued against dues, publications and regional officers for fear that these would compete with the national organization ["Report of...," p. 325]. It has been speculated that the Association was at first concerned that regional groups might form into strong autonomous associations and go their own way in much the same way that the Southern Economic Association had after its disenchantment with the policies of the American Economic Association in the 1920's. [Previts, 1977, p. 23].

This early lukewarm endorsement of regions was followed two decades later by a stronger statement by 1972-1973 president Robert T. Sprouse that:

> Efforts are being made to strengthen the Association's regional organizations in order to facilitate the increased participation that is desired by so many of its members, including the increasing number of academic members from two-year colleges. Many members are rarely, if ever, able to attend the Association's Annual Meeting; regional meetings may present the only reasonable opportunity for professional association with fellow members [Sprouse, 1973, p. 175].

Another event in 1973 was the recognition by the Executive Committee of the need for regional input into the Association's governance process through regional representatives on the national nominating committee. The role of the regions was expanded even more in 1976.

The Southeastern Region was the only regional group until 1957 when the Southwest Region held its first meeting. The Ohio Region held its first meeting in 1958 [Powell, pp. 1-9]. The Northeast and Midwest Regions held their first meetings in 1960, followed by the Western Region in 1966 and the Mid-Atlantic in 1975. There was also a Canadian region from 1967 to 1984, which is discussed in the chapter on "International Outreach Programs."

THE FOUNDING OF THE SOUTHEASTERN REGION OF AAA

The first meeting of the AAA Southeast Region was held at the Georgia Tech Y.M.C.A (on North Avenue, across from the athletic stadium) on April 30, 1949. Lyle E. Campbell of Emory University is credited with promoting and conducting this first meeting [Committe, 1978, p. 4]. He was assisted by a committee composed of S. Paul Garner (University of Alabama), Harold M. Heckman (University of Georgia), Frank S. Kaulback (University of Virginia), and Noah Warren (Georgia Institute of Technology).

The names of fifteen appeared on the program for the 1949 meeting. They were:

Herman Boozer, Georgia Power Company, Atlanta
Lyle E. Campbell, Emory University
Robert L. Dixon (AAA President), University of Michigan
William C. Flewellen, Jr., University of Alabama
L. B. Freeman, North Georgia College
H. M. Heckman, University of Georgia
James S. Lanham, University of Florida
William F. Loflin, CPA, Columbus, Georgia
Robert S. Lynch, Atlantic Steel Co.
George E. Manners, Dean, University of Georgia-Atlanta
Harvey Meyer, University of Tennessee
L. P. Mingledorf, South Georgia College
Paul Thomas, University of Alabama
Noah Warren, Georgia Institute of Technology
William H. Whitney, University of Alabama

Others known to have been in attendance include:

Martin L. Black, Duke University
Everett Bollinger, Georgia Institute of Technology
Joseph Cerny, University of Mississippi
Rube Cooper, Southwestern Publishing Co.
S. Paul Garner, University of Alabama
Frank Kaulback, University of Virginia
Erle Peacock, University of North Carolina
A. J. (Andy) Penz, University of Alabama [Committee, 1978, p. 12].

The influence of the University of Alabama cannot be ignored. Of the 22 names listed above, five were from the University of Alabama. Also, it should be noted that the luncheon speaker was AAA president Robert L. Dixon. This indicates the interest the national organization took in the activities of the new regional group.

Objectives of the Southeastern Region

The initial objectives of the Southeastern Region revolved around holding a professional meeting in a location geographically convenient for members. The recruitment of faculty and learning of placement opportunities were considered important matters at the very beginning. The simple social-contact value of a regional meeting was recognized from the outset [Committe, et. al., 1978, p. 4]. Today, these are essentially the objectives of all of the AAA regional groups. The published objectives of the Southeastern Regional group are somewhat more formalized today in that they are in written form, but a reading of the following illustrates that the regional group's founders had the basic idea in 1949. Today's objectives are:

1. Strive to facilitate a broad-based member participation in the activities and programs of the AAA.
2. Endeavor to sponsor broad-based research and educational programs for the benefit of its members.
3. Offer continuing education programs which shall be coordinated with the AAA's Director of Education.
4. Tailor each specific meeting to the specific needs and desires of the regional members and strive to draw on the talents of a broad spectrum of members in the Region.
5. Publish the proceedings of each meeting [1989-90 Directory, 1989, p. 46].

Regional Officers

Over the years, the Southeast Region has been governed by a committee of five or six individuals. For 1973 and earlier meetings, the head of the committee was simply called the committee chairman. Normally, the chairman was from the host school, but this was not always the case. In 1974, the head of the regional committee was given the designation of regional vice president in an attempt by the AAA Executive Committee to better define the regions' ties with the national organization. From 1974 through 1976 the title was a designation conferred by AAA; in later years an election was held at the annual business meeting. Exhibit 4-1 lists the locations of each of the meetings since the first in 1949, the host school, and the regional chairman (regional vice president beginning in 1974).

The list of committee chairmen reads like a who's who in accounting education. Many of the individuals went on to become national AAA presidents or members of the Executive Committee. The national presidents from the Southeastern Region have included:

Harvey Meyer, 1945, University of Tennessee
S. Paul Garner, 1951, University of Alabama
Willard Graham, 1955, University of North Carolina
Martin L. Black, Jr., 1959, Duke University
Frank Kaulback, 1967-68, University of Virginia,
Lee Brummet, 1974-75, University of North Carolina
Harold Q. Langenderfer, 1983-84, University of North Carolina
John K. Simmons, 1989-90, University of Florida

In addition, James Don Edwards, president in 1970-71 and Herbert E. Miller, president in 1965-66, should be added to the list since they moved to the University of Georgia later in their careers. In fact, Edwards moved to Georgia the year after he was national president. Later, in 1980, Edwards was the regional vice president.

Some individuals were gluttons for punishment in that they agreed to serve as the head of the region for more than one year. These individuals included Harold M. Heckman (1951, 1957, 1964), Robert H. Van Voorhis (1952, 1958), Mervyn W. Wingfield (1967, 1975), and Robert B. Sweeney (1971, 1977). Wingfield was at two different schools, while the others stayed at the same schools for their terms of office.

The Published Proceedings

In addition to being the first recognized regional group of AAA, the Southeastern Region has long boasted (incorrectly) of having the distinction of being the first region to publish the proceedings of the meetings. In 1974, under the leadership of James T. Thompson of Memphis State University, the Memphis State faculty published a proceedings of the papers presented at the meeting. The editors of this first volume were H. Nelson Lunn and L. Gayle Rayburn of the Memphis State faculty. As detailed in the chapter on "International Outreach Programs," the Canadian Region was actually the first to publish its proceedings, beginning in 1969. However, few people attended the Canadian regional meetings, and communication with the national office and the Executive Committee was slight. Consequently, neither the national office nor the Executive Committee apparently knew of the Canadian proceedings. Thus, the proceedings published by other regional groups in subsequent years were pattered after that of the Southeast Region.

The number of authors represented in the proceedings has increased substantially over the years since 1974. A total of 64 authors were represented in the 1974 volume. That number increased to more than 80 authors in the three subsequent

Exhibit 4-1
Southeast AAA Locations and Chief Officers

Year	Location	Regional VP	VP School
1949	Atlanta, GA	L. E. Campbell	Georgia Institute of Technology
1950	Knoxville, TN	Frank S. Kaulback	University of Virginia
1951	Athens, GA	Harold M. Heckman	University of Georgia
1952	Tuscaloosa, AL	Robert H. Van Voorhis	University of Alabama
1953	Chapel Hill, NC	Alton G. Sadler	University of North Carolina
1954	Gainesville, FL	James S. Lanham	University of Florida
1955	Charlottesville, VA	Homer A. Black	University of Georgia
1956	Columbia, SC	E. L. McGowan	University of South Carolina
1957	Athens, GA	Harold M. Heckman	University of Georgia
1958	Baton Rouge, LA	Robert H. Van Voorhis	Louisiana State University
1959	Tallahassee, FL	John E. Champion	Florida State University
1960	Knoxville, TN	A. Weyman Patrick	University of Tennessee
1961	Tuscaloosa, AL	Joseph E. Lane, Jr.	University of Alabama
1962	Durham, NC	Robert L. Dickens	Duke University
1963	New Orleans, LA	Peter Firmin	Tulane University
1964	Athens, GA	Harold M. Heckman	University of Georgia
1965	Gainesville, FL	Williard E. Stone	University of Florida
1966	Chattanooga, TN	Rayford J. McLaurin	University of Chattanooga
1967	Columbia, SC	Mervyn W. Wingfield	University of South Carolina
1968	Atlanta, GA	Catherine E. Miles	Georgia State University
1969	Biloxi, MS	Loomis H. Toler	Mississippi State University
1970	Blacksburg, VA	Ronald J. Patten	Virginia Polytechnic Institute
1971	Tuscaloosa, AL	Robert B. Sweeney	University of Alabama
1972	Baton Rouge, LA	Fritz A. McCameron	Louisiana State University
1973	Chapel Hill, NC	Isaac N. Reynolds	University of North Carolina
1974	Memphis, TN	James T. Thompson	Memphis State University
1975	Richmond, VA	Mervyn W. Wingfield	Virginia Commonwealth University
1976	Tampa, FL	Louis C. Jurgensen	University of South Florida
1977	Nashville, TN	Robert B. Sweeney	University of Alabama
1978	Boone, NC	Floyd Windal	University of Georgia
1979	Miami, FL	Harold Q. Langenderfer	University of North Carolina
1980	Columbia, SC	James Don Edwards	University of Georgia
1981	Birmingham, AL	Pearcy B. Yeargan	University of Alabama-Birmingham
1982	Jacksonville, FL	Roger H. Hermanson	Georgia State University
1983	Virginia Beach, VA	Dora R. Herring	Mississippi State University
1984	Biloxi, MS	Howard P. Sanders	University of South Carolina
1985	Orlando, FL	Keith Bryant, Jr.	University of Alabama-Birmingham
1986	Greenville, SC	James R. Davis	Clemson University
1987	Atlanta, GA	H. Frank Stabler	Georgia State University
1988	Knoxville, TN	Jan R. Williams	University of Tennessee
1989	Washington, D.C.	Larry N. Killough	Virginia Polytechnic Institute
1990	Tampa, FL	Robert J. West	University of South Florida
1991	Birmingham, AL	Robert W. Rouse	Auburn University

years. For the 1979 meeting, the number of authors jumped to 96. The proceedings for the 1980 meeting held in Columbia, South Carolina, contained the work of 125 authors. Another major jump in authorship numbers occurred in 1982 and 1983 when the number increased to 153 individuals. The all-time record of 212 authors came in 1987 in the proceedings of the Atlanta meeting. The numbers dropped off to 153 in 1988, but were back to 189 in 1989. Despite the increased number of papers published, the quality has not diminished over the years. In fact, because of the large numbers of papers submitted, the acceptance rate in recent years has been less than 35 percent. Obviously, it is quite prestigious to have a paper accepted for the Southeastern Regional meeting.

Other regions were quick to copy the idea of publishing a proceedings of each meeting. In fact, the week before the Memphis meeting, the steering committee for the 1975 Mid-Atlantic Regional meeting voted to publish a volume of proceedings for the 1975 meeting. At the time, that group thought they would be the first region to publish proceedings.

Recent Years in the Southeast

In many respects, the history of the Southeastern Region has been less than exciting in recent years. The regional meeting has a long tradition, and each year's meeting seems to follow in the tradition of earlier years. The placement room is extremely popular and professors come from all over the country solely for the purpose of utilizing the placement service. In addition, the regional meeting is popular with publishers. The publishers have the opportunity to show their wares to a large number of professors at a time when the professors are still in the process of making textbook decisions for the following academic year. The publishers have been very supportive of the regional activities with companies such as Irwin and South-Western sponsoring cocktail parties and breakfasts. South-Western Publishing Co. has been represented at every meeting. That firm's representative, Hal Cole, of Charlotte, North Carolina, attended 36 consecutive meetings beginning in 1950. The publisher exhibits are indeed an important part of the meetings. In fact, the popularity of the book exhibit area and placement service has been such that many individuals spend more time in those areas than at the formal presentations.

Most meetings in earlier years were held on university campuses. However, with increasing attendance, it became necessary to move the meetings to large hotels. The 1978 meeting held in Boone, North Carolina was the last to be held on a university campus, and even that required attendees to be housed at some 30 motels in the Boone area. Since then, all meetings have been held in large city hotels, although some have been held close to campus such as the one in Columbia, South Carolina in 1980.

One change that has been noticeable during the past decade has been the increase in activity on the Thursday before the conference actually begins. In 1977, the only activity that took place on Thursday was conference registration. Today, there are a variety of continuing-education-type programs . In addition, the Southeast Graduate Workshop, recently called the Southeast Doctoral Consortium, meets on the day before the conference. This group provides an opportunity for doctoral students to meet with research faculty members from around the region.

As the first regional group within the AAA structure, the Southeastern Region has had a major impact on the regionalization of the Association. In addition, the Southeastern Region has provided a large number of national officers. In a sense, the regional activities have served as a training ground for serving AAA at the national level. Although the region is no longer the largest in terms of membership (the Western Region has about 1,500 members to 1,400 for the Southeastern Region), it still boasts the largest attendance of any of the regional meetings.

THE SOUTHWESTERN REGION

From documents available at the AAA headquarters in Sarasota, it appears that the organizers of the first Southwest Regional meeting were Professors James M. Owen (Louisiana State University), William F. Crum (Wichita State University), Emerson O. Henke (Baylor University), Vernon H. Upchurch (University of Oklahoma), and Nolan E. Williams (University of Arkansas). Fred Norwood, then with Peat, Marwick, Mitchell & Co. (later with Texas Tech University), was the first regional chairman.

From the beginning through the 1973 meeting, the Southwest Region of AAA met jointly with the Southwestern Social Science Association annual convention. In 1974, the group began meeting with the Southwest Federation of Administrative Disciplines (SWFAD). Joint meetings with these other groups was viewed as desirable because at many small schools the faculty members had teaching responsibilities in more than one subject area. The joint meetings permitted professors to attend sessions in all subject areas in which they taught.

The switch to SWFAD in 1974 involved a mass exodus of all of the business disciplines from the Southwestern Social Science Association. The original seed for a SWFAD-type organization was planted by Bob Booser, then with Richard D. Irwin, Inc. Booser, during a conversation with Dennis Ford (Dean at Texas A & I), objected to the social scientists who came to his publisher's cocktail parties even though they were not interested in business textbooks. Booser believed that the business disciplines were large enough to have their own organization [Starling and Bruno, 1990, p. 44-45]. Later, Ford became president of the Southwest Business Administration Association and was able to convince others of the merits of a separate organization for business groups. At the same time, there was discontentment within some of the organizations with respect to the facilities they were being assigned by the social science group. As a result of this discontent, the Academy of Management voted to hold its 1973 meeting in the same city at the same time as the social science group, but at a different hotel. Henry Hays, president of the Academy of Management, met with the president of each of the business disciplines and asked them to form a new business organization. As a result of these meetings, the officers of the business disciplines composed a declaration of independence which was submitted to the Southwest Social Science Association on March 24, 1973. James Modisette of the University of Arkansas was the accounting representative who signed the declaration [Starling and Bruno, 1990, p. 46]. The first business meeting of SWFAD officers was held on April 8, 1973, at the Ramada Inn near Love Field in Dallas. Robert L. Grinaker (University of Houston) represented the AAA [Starling and Bruno, 1990, p. 47]. Most years the AAA has been the largest group represented at SWFAD meetings, however on a few occasions the Academy of Management has had a few more registrants than the accountants.

In 1975, there was consideration of having the AAA withdraw from SWFAD and meet independently. Robert Grinaker held a meeting at the Houston airport on the day before the 1975 SWFAD meeting of 18 department chairmen from throughout the region. The 18 chairmen voted unanimously to recommend that the Southwest AAA meeting be held separately from SWFAD in future years. That recommendation was made at the formal Southwest AAA business meeting the next day, but the membership defeated the motion by one vote [Hood, 1990]. Again in 1990 a committee was formed to evaluate a similar proposal.

Southwestern Program Activities

The extent of activities was not great in the early years of the Southwestern Region. The technical program for the 1959 meeting in Galveston included six papers. On Friday, Fred Norwood of Texas Tech presented a paper entitled "Management Services by Certified Public Accountants;" Othel D. Westfall of the University of Oklahoma spoke on "Linear Programming and the Accountant." A Friday afternoon paper presented by Roger L. Holmes of Baylor University, was entitled "Social Responsibility of the Certified Public Accountant." Jim Ashburne of the University of Texas spoke on "Cooperation with the Accounting Profession." On Saturday morning, O. J. Curry of North Texas State College presented a paper entitled "A Dean's Look at the Accounting Department." This was followed by a presentation by Walter Plumhoff of Arthur Andersen & Co. entitled "The Practitioner Looks at the College Accounting Program" ["AAA Regional...," 1960, p. 363]. The 1960 meeting in Dallas was similar to that of the preceding year, but included an address by AAA president Charles Gaa (Michigan State University).

The 1961 meeting extended over a day-and-a-half, but there were only three papers presented by academicians (Glenn Welsch of Texas, Roderick Holmes of Baylor, and I. E. McNeill of the University of Houston). There was also a report on AAA activities by Association president Ben Carson (UCLA). The remainder of the time was filled by three presentations by practitioners and three panel discussions. The most noticeable difference between these early meetings and those of today was the heavy inclusion of practitioners from both public accounting and industry on the program.

By 1969, there were six accounting sessions, most with one speaker, and this included the president-elect's (Norton Bedford) speech. However, by 1973 (the last year of meeting with the Southwestern Social Science Association), the conference had expanded to a day and a half with 16 speakers (including president elect Robert Anthony) addressing 11 topics.

The first year of meeting with SWFAD was little different from the year before. The 1974 meeting, held at the Hotel Adolphus in Dallas, began on Thursday with a plenary session headed by Guy W. Trump of the AICPA, who spoke on professional schools of accounting. Four concurrent sessions filled up the remainder of the afternoon. Another plenary session was held on Friday morning followed by two concurrent sessions and a panel discussion. The title of one of the concurrent sessions was "Current Research in Accounting: University of Arkansas." Four Arkansas professors delivered two papers. The two discussants and the moderator were also from the University of Arkansas. The business meeting, beginning at 2:30 on Friday afternoon, concluded the meeting. Altogether, 18 individuals presented a total of 13 papers. Six of the individuals were from the University of Texas, four from the University of Arkansas, three from North Texas State, and two from Oklahoma State University.

The 1975 meeting in Houston, led by regional vice president Kenneth Most of Texas A & M University, saw a marked change in the level of program participation. Kenneth Most was more ambitious than his predecessors with respect to program development. The meeting covered two full days—beginning Thursday morning with two separate plenary sessions (headed by Robert Sterling of Rice University and Harvey Kapnick of Arthur Andersen & Co.). This was followed in the afternoon by six concurrent sessions. For the first time, there were three concurrent sessions held at a time to give attendees a wider choice of sessions to attend. Each session included three papers—again, an expansion from previous years. Altogether, 72 individuals presented papers at the 1975 meeting—a four-fold increase over the preceding year. In addition to the speakers, there were many discussants and moderators who also appeared on the program. AAA president-elect Wilton Anderson (Oklahoma State University) was the luncheon speaker. A similar number of program participants were to appear on the program of the 1976 meeting. The heavy participation continued to increase in the late 1970s and throughout the 1980s.

Recently, Kenneth Most stated that he had first become active in regional meetings while living in the Southeast Region. The annual meeting in that region has long been a participative one with a high percentage of the attendees being active in the program. Therefore, Dr. Most tried to pattern the Southwest meeting after the meetings in the Southeast Region. His experiment was most successful and has resulted in a high level of participation in all meetings since 1975 [Most, 1990].

By 1981, an average of about 70 papers were being presented each year (although this number declined to 42 in 1985). There were 66 papers presented at the 1989 meeting by 228 program participants (just 16 short of the record number of participants in 1984). The 1989 attendance was 327, just four short of the 1988 record of 331.

Regional Officers

In the early years, one individual served as the head of the region, holding the title of regional chairman, and later, regional vice president, and as program coordinator. That pattern was broken in 1976 when Robert Sterling (Rice University) was elected vice president and Allen Porter was named program coordinator. Porter became regional vice president three years later. From 1979 through 1981, a pattern emerged of having the regional vice president be an individual who had previously served as the program coordinator. However, in 1980, program director Ruth bullard (University of Texas at San Antonio) was not nominated for regional vice president because some members of the nominating committee felt that nominees should be well known academicians at leading institutions. Many members, however, thought the highest office should be a reward for serving as program chairperson [Jones, 1990]. Thus, Bullard was nominated from the floor because a petition requesting her nomination had been circulated and signed by numerous members. She was elected over the candidate selected by the nominating committee. At the 1981 meeting the program director, Richard Jones of Lamar University, was not nominated by the nominating committee for the position of 1982 regional vice president. However, he was nominated from the floor by Ruth Bullard. Following the nominating speeches and some questioning of the legitimacy of such a nomination, the number of people in attendance had dwindled to about fifteen. When the vote was taken, Jones won the election and preserved the tradition of the program coordinator continuing on as regional vice president the following year. That tradition has subsequently been broken twice, in 1985 and 1988, when individuals could not serve. A complete list of the regional vice presidents and program coordinators, along with the location of each meeting appears in Exhibit 4-2.

Despite its small size, the Southwestern Region has been a prolific source of individuals who have served AAA at the national level. In particular, the region has

Exhibit 4-2

Southwest Region Meeting Locations and Officers

Year	Location	Regional VP	Program Coordinator
1957		Fred W. Norwood, Peat, Marwick, Mitchell & Co.	
1958		James M. Owen, Louisiana State University	
1959	Galveston, TX	Emerson O. Henke, Baylor University	
1960	Dallas, TX	William P. Carr, Loyola University-New Orleans	
1961	Dallas, TX	Horace R. Brock, North Texas State College	
1962	Dallas, TX	J. Herman Brasseaux, Louisiana State U. at New Orleans	
1963	San Antonio, TX	James T. Johnson, Louisiana Polytechnic Institute	
1964	Dallas, TX	I. E. McNeil, University of Houston	
1965	Dallas, TX	Roderick L. Holmes, Baylor University	
1966	New Orleans, LA	James T. Hood, University of Texas	
1967	Dallas, TX	Barry G. King, Oklahoma State University	
1968	Dallas, TX	Bernard A. Coda, North Texas State University	
1969	Houston, TX	Edward L. Summers, University of Texas-Austin	
1970	Dallas, TX	Howard J. Snavely, University of Texas-Arlington	
1971	Dallas, TX	Milton Usry, Oklahoma State University	
1972	San Antonio, TX	Frank J. Imke, Texas Tech College	
1973	Dallas, TX	James P. Modisett, University of Arkansas	
1974	Dallas, TX	Robert Grinaker, University of Houston	
1975	Houston, TX	Kenneth S. Most, Texas A&M University	
1976	San Antonio, TX	Robert R. Sterling, Rice University	Alen Porter
1977	New Orleans, LA	Kermit D. Larson, University of Texas-Austin	W. Morris/T. Klammer
1978	Dallas, TX	Doyle Williams, Texas Tech University	H. Vaden Streetman
1979	Houston, TX	Alan Porter, University of Tulsa	Paul A. Dierks
1980	San Antonio, TX	Paul A. Dierks, University of Texas-Arlington	Ruth H. Bullard
1981	New Orleans, LA	Ruth H. Bullard, University of Texas-San Antonio	Richard W. Jones
1982	Dallas, TX	Richard W. Jones, Lamar University	Ted L. Fisher
1983	Houston, TX	Ted L. Fisher, Louisiana Tech University	E. Hawkins/C. Sangster
1984	San Antonio, TX	Ennis M. Hawkins, San Houston State University	Sam Sedki
1985	New Orleans, LA	Horace R. Brock, North Texas State University	Michael F. Foran
1986	Dallas, TX	Sam Sedki, St. Mary's University	Jimie Kusel
1987	Houston, TX	Jimie Kusel, University of Arkansas-Little Rock	Bruce Swindle
1988	San Antonio, TX	Bruce Swindle, McNeese State University	Jack R. Ethridge
1989	New Orleans, LA	Sammie L. Smith, Stephen F. Austin State	S. Wescot/D. Finley
1990	Dallas, TX	David R. Finley, University of Houston	Clifford E. Hutton
1991	Houston, TX	Clifford E. Hutton, University of Tulsa	Jackson A. White

supplied five (or six, depending upon how you count) national presidents since 1957: Glenn Welsch (University of Texas), Charles T. Zlatkovich (University of Texas), Wilton Anderson (Oklahoma State University), Stephen Zeff (Rice University), and Ray Sommerfeld (University of Texas). Furthermore, the region can claim Doyle Williams as another of its contributions to the office of president. Williams served as the regional vice president in 1978 while at Texas Tech University. That same year he moved to the University of Southern California and later became national president in 1984. The region has also furnished three editors of *The Accounting Review*, namely Charles Griffin (University of Texas), Stephen Zeff (Rice University), and William Kinney (University of Texas).

Southwestern Region Summary

The Southwestern Region was the second regional group formed within the American Accounting Association. The group has had a dynamic history over its third-of-a-century existence. Key years in the group's history besides the 1957 initial meeting include the 1974 switch to meeting with SWFAD and the 1975 increase in program participation. Even though the number of program participants is far greater today than was the case in the early years, it is noticeable that there has been little change in program subject matter. The titles of the papers presented in the late 1950s and early 1960s would be appropriate today.   In some respects, the region is different from the other regional groups of the AAA. For instance, holding its meetings jointly with other groups is a characteristic shared with no other region.   In addition, the annual meetings of the region are heavily attended by individuals from outside the region. Thus, the meetings seem in many respects similar to a national meeting rather than a regional meeting. Given the small size of the regional membership, this interest from professors outside the region contributes greatly to what might otherwise become a provincial meeting. In recent years there have been a little over 300 accounting registrants out of the approximately 1,800 SWFAD attendees. The region, although one of the smallest in terms of academic members, has provided more than its share of national officers.

THE OHIO REGION

The Ohio Region is unique among AAA regions in that it is the only region composed of members in only one state. This provinciality has at times caused recognition problems within the Association's hierarchy. The Ohio regional group grew out of a state-wide group of accounting educators that had been meeting regularly in Columbus. The 1957 meeting of that group was attended by AAA president C. A. Moyer (University of Illinois) and a request was presented to him asking for AAA recognition. The first meeting held as an AAA region was in December 1958 in Columbus. The 1958 AAA president, C. Rollin Niswonger (Miami University), attended that first meeting [Powell, 1974, p. 3] as did fifty-eight faculty members from nineteen Ohio institutions. With the exception of Richard M. Cyert of Carnegie Institute of Technology, all of the program participants were from Ohio colleges and universities [Epaves and McKeon, 1984, p. 4]. The meeting lasted only one day, as did subsequent meetings through 1968.

The 1959 meeting, also held in Columbus, was attended by 90 educators from 27 colleges and universities. The 1959 program consisted primarily of panel discussions. All participants were from the state of Ohio. The entire state board of accountancy participated in the program as did the board's Executive Secretary [Epaves and McKeon, 1984, p. 5].

Throughout the 1960's all meetings were held in Columbus except for 1969 when Cincinnati hosted the region. Through 1963, the meetings were held in December—which in Ohio can mean snow. Meetings were not held in 1961 and 1962 (although the 1963 meeting was called the 1962 meeting). Snow caused the 1961 meeting to be cancelled and the 1962 meeting to be postponed until February 1963. At that point, it was decided to move the meetings to the spring. Therefore, since a meeting had already been held in February 1963, and December 1963 would be only four months before the April 1964 meeting, it was decided to cancel the December 1963 meeting. All meetings have been held in April or May since 1964 [Epaves and McKeon, 1984, p. 6].

The 1966 meeting was held as a joint meeting with the Midwest Region. E. Ben Yager of Miami University was co-chairman with Victor H. Tidwell of Kansas State University. This joint meeting may have been a political mistake, however, because the AAA Executive Committee began questioning whether Ohio was merely a state within the Midwest Region, or a separate region on its own behalf. The 1966 Long-Range Planning Committee included Ohio as a member of the Midwest Region. The 1969 Committee on Regional Arrangements had also included Ohio in the Midwest Region. By 1969, the Midwest Region had 2,060 members, excluding Ohio. Ohio had 590 members—making it the smallest regional group other than Canada (next smallest was the Southwestern Region with 1,010 members) [Gerhardt, November 3, 1969]. Gerhardt supported the idea of keeping Ohio separate because he felt that the smaller groups would encourage participation by all members. In 1973, however, there was still question as to whether Ohio was a state or a region. The minutes of the August 1973 Executive Committee state that a decision must be made because regional subsidies based on Ohio members should be sent to the Midwest Region if Ohio were not a region. It was agreed that Ohio thought it was a region.

The University of Cincinnati hosted the 1969 meeting in its home city. This was the first two-day meeting of the region. This began the tradition of allowing universities other than Ohio State University to host the meetings. Several other schools hosted the meetings held during the 1970s.

The meetings of the Ohio Regional Group became less provincial during the 1970s. In addition to the AAA presidents-elect, there were numerous speakers from outside of the region. Philip Defliese (Lybrand, Ross Bros. & Montgomery) was the key speaker at the 1972 meeting. Abraham Briloff (City University of New York) was the main drawing card at the 1974 meeting. Other big names from outside the region who spoke during the 1970s and early 1980s (other than the AAA president-elect) included Harvey Kapnick (Arthur Andersen & Co.), Robert Sweeney (University of Alabama), Robert Grinaker (University of Houston), James MacNeill (AICPA), Richard Brief (New York University), Yuji Ijiri (Carnegie Mellon), and Harold Langenderfer

Exhibit 4-3

Ohio Region Officers and Meeting Locations

Year	Location	Chair/VP	Program Chair
1958	Columbus	James R. McCoy, University of Akron	
1959	Columbus	Dennis Gordon, University of Akron	
1960	Columbus	Ralph F. Beckert, Ohio University	
1961	No Meeting	Norwood C. Geis, University of Cincinnati	
1962	Columbus	Glen G. Yankee, Miami University	
1963	No Meeting	Glen G. Yankee, Western Reserve University	
1964	Columbus	Edwin C. Bomeli, Bowling Green State University	
1965	Columbus	Francis J. McGurr, John Carroll University	
1966	Columbus	E. Ben Yager-Co-Chair, Miami	
1967	Columbus	Donald F. Pabst, University of Cincinnati	
1968	Columbus	Henri C. Pusker, Kent State University	
1969	Cincinnati	Richard S. Roberts, University of Akron	
1970	Akron	Vincent M. Panichi, John Carroll University	
1971	Columbus	Ronald V. Hartley, Bowling Green State University	
1972	Toledo	Charles F. Story, Muskingum College	
1973	Kent	Robert E. Boggs, Miami University	
1974	Dayton	Donald F. Pabst, Wright State University	Clara C. Lelievre
1975	Bowling Green	Wayne Johnson, Bowling Green State University	Nabil Hassan
1976	Athens	C. R. Stevenson, Ohio University	Wayne Johnson
1977	Columbus	Charles Gibson, University of Toledo	Hal Jasper
1978	Cincinnati	Clara C. Lelievre, University of Cincinnati	Felix Kollaritsch
1979	Columbus	Robert T. Sullens, John Carroll University	Charles Carpenter
1980	Cleveland	Richard A. Epaves, Cleveland State University	Lawrence C. Phillips
1981	Columbus	Richard J. Murdock, Ohio State University	William Voss
1982	Toledo	Blaine A. Ritts, Bowling Green State University	Robert T. Sullens
1983	Columbus	Ray G. Stephens, Ohio State University	Robert T. Sullens
1984	Oxford	Gary J. Previts, Case Western Reserve University	John Cumming
1985	Columbus	Elise G. Jancura, Cleveland State University	Bruce R. Gaumnitz
1986	Akron	Daniel L. Jensen, Ohio State University	Dennis Lee Kimmell
1987	Columbus	Charles H. Gibson, University of Toledo	Thomas J. Burns
1988	Cincinnati	Blaine A. Ritts, Bowling Green State University	J. Timothy Sale
1989	Columbus	Richard J. Murdock, Ohio State University	Richard W. Metcalf
1990	Perrysburg	John Cumming, Miami University	Joyce S. Allen
1991	Cleveland	Ray G. Stephens, Ohio State University	Jane E. Campbell

(North Carolina). By 1980, over one-third of the concurrent session papers were authored by professors from outside of the region. Outsiders had become a majority by 1983 [Epaves and McKeon, 1984, p. 11].

Ohio Regional Officers

Exhibit 4-3 lists the locations of the regional meetings and the regional chairpersons and vice presidents. Six individuals have headed the region for two terms: Clara C. Lelievre, Wayne Johnson, Robert Sullens, Donald Pabst, Ray Stephens, and Glen Yankee. Although none of the regional officers have yet gone on to become

AAA national presidents, the regional officers have progressed in the AAA hierarchy. Charles Carpenter (Miami University) became a national vice president, while Thomas Burns (Ohio State) served as director of education. Other individuals have served as committee chairpersons.

The affiliation of the regions with the national office of AAA was somewhat loose during the 1960s, and this was a major concern of some people—particularly administrative secretary Paul Gerhardt. Gerhardt noted in a 1969 letter that president Norton Bedford had received a list of the members of the Ohio regional committee and had gone through the formality of writing a letter officially appointing these individuals. However, when Gerhardt checked the membership records, three out of the seven committee members were not members of the AAA. Gerhardt felt that this was an example of too much independence by the regions [Gerhardt, November 3, 1969].

The Ohio Region has been surprisingly prosperous given its small size. Historically, it has been the smallest region and most provincial, but meeting attendance has been quite high as a percentage of total membership within the state. Also, the region has seemingly served as a training ground for individuals going on to more important positions within the AAA structure. Although a region of such small size could certainly be considered a questionable experiment, it has proven to be successful.

THE NORTHEAST REGION

The Northeast Regional group of the AAA held its first meeting in 1959, but this was essentially an organizational meeting. According to AAA records, the first official meeting of the Northeast Region was in 1960. The Northeast Region traces its roots back to a 1951 meeting of the Conference of New England Accounting Instructors which was held on the campus of the Massachusetts Institute of Technology [Wood, 1977, p. 64].

On April 25, 1959, a planning meeting was held at College of the Holy Cross. The individuals responsible for calling this meeting were John W. Anderson and Robert W. Lentilhon of the University of Massachusetts and Reginald J. Smith and John O'Connell of Holy Cross. Invitations were sent to colleges in New England, New York, and New Jersey asking each school to send one representative to the meeting. Thirty-one individuals actually made it to the meeting. It was decided to hold a regional meeting in the fall of 1959 and to affiliate with the American Accounting Association. Three institutions offered to host the first meeting—Stonehill College, Northeastern University, and Hofstra University. Hofstra won on a second ballot by a vote of 22 to 9. It was decided that the meetings should be rotated between the Boston and New York areas [Wood, 1977, p. 65]. The first meeting was held at Hofstra on November 13 and 14, 1959. At that time the region consisted of all of New England, New York, New Jersey, Pennsylvania, and Southeastern Canada. In later years, the region was reduced to only New England and New York.

The 1960 meeting committee was chaired by Gordon Shillinglaw, then at MIT. The big controversey at that meeting was whether New Yorkers should be allowed to participate in the region. The proper Bostonians wanted the membership to be restricted solely to New England. Shillinglaw remembers being responsible for preparing the program and selecting the speakers. The program brochure was merely a mimeographed sheet. Shillinglaw recalls that the acrimony over membership stemmed largely from the views of some—particularly Robert Anthony of Harvard—that the 1951 meeting was quite nice in that it was limited to New Englanders. Robert Lentilhon was criticized for unilaterally naming some New Yorkers to the organizational committee [Shillinglaw, 1990].

The Northeast Regional meeting has grown into a large program in recent years. The 1969 attendance was 263, and that minimum was maintained throughout the 1970s (even after the 1975 split-off of the Mid-Atlantic Region caused the loss of Pennsylvania and New Jersey from the region). As an indication of program size, there were 65 program participants in 1973. That number increased to 95 in 1974. A highlight in the history of program booklets came in 1980 when Norm Berman published a 72-page volume—the largest in the region's history. Detailed biographies were provided for every speaker.

Northeast Region Governance

In the early years, a committee of five individuals planned and organized the regional meetings, in a manner similar to that of the Southeastern Region. However, by 1970, the committee structure had fallen by the wayside and the meeting was the sole responsibility of the host school. It was in 1970 that Porter Wood of the University of Rhode Island (the host of the 1971 meeting) reorganized the region and established a nine-person steering committee to coordinate future regional meetings. Wood was unaware that a previous five-member committee structure had ever existed. The steering committee provided for greater continuity between the annual meetings.

Exhibit 4-4 is a list of the sites of all of the regional meetings and the regional officers. Until the mid-1970's, the committee chairperson or regional vice president also served as the program coordinator. In more recent years, the position of program coordinator has been the responsibility of the vice-president-elect. One noticeable difference between the regional leaders of the Northeast Region and those of other regions is that the Northeast leaders have rarely gone on to higher positions within

Exhibit 4-4
Northeast Region

Year	Location	Regional VP	Program Coor.
1959	Hempstead, NY	William H. Childs, Hofstra College	
1959	Cambridge, MA	Gordon Shillinglaw, Mass. Institute of Technology	
1961	Jamaica, NY	R. V. Lucano, St. John's University	
1962	Boston, MA	Richard Vancil, Harvard University	
1963	No Meeting		
1964	New York, NY	Michael Schiff, New York University	
1965	Babson Park, MA	Clinton A. Petersen, Babson Institute	
1966	New York, NY	Gordon Shillinglaw, Columbia University	
1967	Boston, MA	Raymond L. Mannix, Boston University	
1968	Hempstead, NY	James A. Cashin, Hofstra University	
1969	Waltham, MA	Rae D. Anderson, Bentley College	
1970	New York, NY	Henry H. Bolz, Fordham University	
1971	Kingston, RI	Porter S. Wood, University of Rhode Island	
1972	Garden City, NY	Lee Seidler, New York University	
1973	Amherst, MA	Carl Dennler, Jr., University of Mass.-Amherst	
1974	Philadelphia, PA	Bruce Oliver, University of Pennsylvania	
1975	Albany, NY	Donald F. Arnold, SUNY-Albany	
1976	Boston, MA	Carl Dennler, University of Massachusetts	Paul Janell
1977	New York, NY	Robert L. Gray, New York Society of CPAs	Robert L. Gray
1978	Hartford, CT	Donald Arnold, SUNY-Albany	
1979	Amherst, MA	Paul A. Janell, Northeastern University	Ula Motekat
1980	New York, NY	Martin Benis, Baruch College, City University	N. D. Berman
1981	Boston, MA	George H. Sorter, New York University	Michael J. Sandretto
1982	Providence, RI	Michael J. Sandretto, Harvard University	Spencer J. Martin
1983	New York, NY	Spencer J. Martin, University of Rhode Island	Bernard H. Newman
1984	Boston, MA	Bernard H. Newman, Pace University	Descom D. Hoagland
1985	Syracuse, NY	Descom D. Hoagland, III, Babson College	Mohamed Onsi
1986	Newport, RI	Mohamed Onsi, Syracuse University	Spencer Martin
1987	Hartford, CT	Spencer Martin, University of Rhode Island	Corine T. Norgaard
1988	Burlington, VT	Corine T. Norgaard, University of Connecticut	Peter E. Battelle
1989	Albany, NY	Peter E. Battelle, University of Vermont	Donald F. Arnold
1990	New York, NY	Donald F. Arnold, Union College	Samir B. Fahmy
1991	Springfield, MA	Samir Fahmy, St. Johns University	Anthony T. Krzystofik

the national AAA structure. Whereas serving as head of the Southeast or Southwestern Region seemed to be a stepping stone, such has not been the case for Northeast Region leaders. In fact, there has never been a Northeast Region chairman or vice president go on to become national president. Since the founding of the regional group, there have been only two presidents from the region—Robert Anthony (Harvard) in 1973-74 and Thomas Dyckman (Cornell) in 1981-82. Of these, Robert Anthony was somewhat active in regional activities in the early days.

Another oddity about the Northeast Region is the lack of published proceedings. Unlike the other regions, the Northeast did not publish proceedings until 1990. However, the Region did try to publish a journal, *The Accounting Journal*, in 1977 and 1978. Unfortunately, that effort was short lived. The lack of published proceedings is somewhat unusual given the large size of the region. Attendance has stayed in the area of 250 to a little over 300 registrants every year from the late 1960s to the present time. In most years, this attendance was greater than that at meetings of the Midwest and Mid-Atlantic Regions—both of which did publish proceedings.

One of the more interesting footnotes in the history of the region was in the early 1980's when there was a movement to have Puerto Rico admitted to the region. Since regional meetings must be held within the region, the Northeast steering committee thought that it would be a good idea to be able to meet occasionally in the Caribbean. Nothing came of the matter and Puerto Rico is still not affiliated with any region. At that time, there were fewer than 20 AAA members living in Puerto Rico.

In summary, the Northeast Regional Group has been quite successful over the years, but it has perhaps been more teaching oriented than some of the other regions. From the beginning it was felt that there were enough professional meetings where theoretical papers could be presented and therefore the Northeast Regional meeting should be more directed at the problems facing the teacher. As a result, there has been intense interest from instructors at small colleges and universities [Wood, 1977, p. 65]. In fact, many of the host institutions have been the small schools of the region—such schools as Fordham, Hofstra, Babson, Bentley, Adelphi, and Union College. Perhaps it is this emphasis on the problems of the smaller schools that helps explain why regional service has not been the stepping stone to greater things in the same manner that has been typical in other regions.

THE MIDWEST REGION

The Midwest Region is today the largest of the American Accounting Association (AAA) regions. Included are members in the states of Indiana, Illinois, Iowa, Kansas, Michigan, Missouri, Minnesota, Nebraska, North Dakota, South Dakota, and Wisconsin. Many of the states on the Western edge of the Region, however, are several hundred miles from the normal meeting sites of the annual meeting. Consequently, the region has not always realized its full potential given its large numbers. Those membership numbers have varied over the years depending upon which states were included in the region. Prior to 1973, the Dakotas, Nebraska, Missouri, and Kansas were not considered a part of the Midwest Region (they were in the nonexistent Great Plains Region). Also, some lists prepared by the AAA over the years have erroneously included Ohio as a part of the region. In fact, the Ohio Region and the Midwest Region held a joint annual meeting in 1966, in Columbus, Ohio.

The Midwest Region, like the Southwest Region, for many years held its annual meeting in association with another group. In the early years, the group was the

Midwest Economics Association. In 1965, the Midwest Business Administration Association (MBAA) was established and, beginning with the 1965 meeting, the AAA met in association with that group. Finally, the accountants decided to split off from the other groups and go it alone. The 1987 meeting, held in Milwaukee without the MBAA, was a huge success as it drew the largest attendance ever for a Midwest Region meeting ["1987 Midwest...," 1986, p. 18].

The Early Years

The AAA Midwest Region was approved at the August 1959 AAA Executive Committee meeting in Boulder. Actually, the regional group had met for several years as the accounting section of the Midwest Economics Association, but it was not until the 1959 meeting that members voted to request recognition from the AAA. James Don Edwards, then of Michigan State University, was the program chairman for the 1959 meeting and it was he who pushed for affiliation with the American Accounting Association. Academic speakers at the 1959 meeting consisted of Gardner Jones (Michigan State University) and Sidney Davidson (University of Chicago) ["AAA Regional...," 1960, p. 365]. James Don Edwards nominated Leon Hay (Indiana University) to serve as the first AAA Midwest Regional chairman. There were no other nominations and Hay was elected by acclamation [Hay, 1990].

The 1960 meeting in Minneapolis, the first under the auspices of the AAA, was organized by Leon Hay, who expanded the program slightly from that of preceding years. Other members of the first regional committee included Donald W. Brown (Iowa State University), Gardner Jones (Michigan State University), R. D. Koppenhaver (University of North Dakota), and Stephen W. Vasquez (St. Louis University). Speakers for the 1960 meeting included Ray Powell (Notre Dame) who spoke on "Depreciation Reform," Raymond C. Dein (University of Nebraska), who spoke on "Responsibilities in Accounting Research," and William Ferrara (University of Illinois), whose topic was "The Structure of Cost Control." There was also a panel discussion entitled "Are College Accounting Curricula Unfairly Dominated by CPA Examination Requirements?" ["Midwestern...," 1960, p. 758]. With a topic like that, it sounds as if little has changed in accounting education over the past three decades. Attendance was not particularly high at the early meetings. Joe DeMaris recalls that only twelve attended the 1961 meeting in Indianapolis.

The 1966 meeting of the Midwest Business Administration Association was held in Columbus, Ohio, on the campus of Ohio State University. This meant that the Midwest Region of AAA had to meet outside of its region. Actually, a joint meeting was held with the Ohio Region.

Until 1974, the program format of the Midwest Region meeting contained a plenary session for the main speaker and discussants, and approximately four additional sessions for panels or speakers. Richard Ortman, 1974 Vice President, instituted the use of numerous concurrent sessions in the program format. This change provided a diversified program in terms of accounting research topics, and increased the number of participants to about 30 people. The 1975 program committee felt that wide participation from members of the region was desirable, and retained the program format of the prior year. Unfortunately, the 1975 attendance was severely hurt by a heavy snow which fell on Chicago on the first day of the meeting.

A volume of proceedings was first published in 1975. The proceedings had two purposes: (1) a formal recognition of the effort expended by participants to enhance accounting research; and (2) a documented reference for use by interested parties subsequent to the annual meeting. The initial volume was funded by the Karl M. Doeren Foundation, through the assistance of James J. Leisenring, a partner in the firm of Doeren, Mayhew, Grob & McNamara. The editor of the first volume was John D. Sheppard (Western Michigan University). The Proceedings were published in August following the annual meeting in April.

John D. Sheppard, 1974-75 vice president, submitted a most impressive report including highlights of the annual meeting and recommendations for the following year, including: Achieve financial stability for the region by increasing the registration fee to cover costs for the annual program. At this time the AAA was not charging a separate fee; all registration fees (which were only $2) went to the MBAA. He also recommended the initiation of pre-meeting professional development courses and increased communications with MBAA to improve publicity for the meeting. There had been two professional development seminars at the 1974 meeting, one dealing with applications of the behavioral sciences to auditing and the other applying facets of behavioral sciences to managerial accounting.

A total of 25 papers was presented at the 1976 meeting in St. Louis. These consisted of 18 papers presented by faculty members or practitioners at concurrent sessions, three papers presented by students at a concurrent session, and four plenary session papers. Only five of the papers were coauthored [Buzby, 1976, p. iii]. All of the authors were from within the region except for plenary speaker Robert Anthony (Harvard University), a past AAA president, three Canadians, and a student from Ohio State University. It should be noted that the AAA Executive Committee had, in 1966, issued guidelines which suggested that speakers at regional group meetings should be selected from the geographical region "represented by the group to the fullest extent possible, in order to encourage and develop such regional talent" [Minutes, 1966, p. 12].

The size of the program had increased substantially by 1981 when the group met at the Palmer House in Chicago. There was even a four-hour continuing education program on international accounting on the day prior to the meeting. The international accounting program was conducted by Hanns-Martin Schoenfeld (University of Illinois) and James Gaertner (Notre Dame). The 1981 meeting featured 41 accounting papers, written by 67 authors. Apparently the 1966 policy recommending that speakers come from within the region has been forgotten by this time. Of the 67 authors at the 1981 meeting, 26 were from institutions outside of the Midwest Region. This heavy contribution from individuals outside of the region was to continue in future years; the Midwest Regional had become a national meeting.

Three years later, the program for the 1984 meeting in Chicago had grown substantially. An analysis of the 1984 program book shows that there were 64 accounting papers presented by 105 authors. There were also five continuing education programs on the day prior to the beginning of the annual meeting. Fully one-third, or 35, of the authors were from outside of the Midwest Region. Outsiders came from as far as California, Florida, and the New England states. The Midwest Regional meeting had indeed become a national meeting. Interestingly, there were no Canadian speakers in 1984—the group most represented among the outsiders in 1976.

The long discussed first meeting separate from the MBAA took place in 1987 at the Pfister Hotel in Milwaukee. The attendance of 311 made for a successful meeting. There were also 17 book publishers exhibiting their wares. There were five continuing education sessions on the day prior to the meeting. Over 80 papers were presented at 30 concurrent paper sessions.

The number of premeeting continuing education programs had increased to six by 1988 and there was a dinner for manuscript authors. There were 75 papers presented, written by 114 authors. Fifty of the authors were from outside of the region. In addition, there were ten sessions made up of panel discussions with a total of 34 panelists. Only five of the panelists were from outside of the region. AAA President-Elect Gerhard Mueller was a breakfast speaker. The program booklet for the 1988 meeting lists 55 individuals who had served on the Manuscripts Review Committee. In 1976, all papers had been reviewed by the nine members of the program committee.

As mentioned previously, the proceedings of the regional meeting were first published in 1975. Such publication continued in subsequent years, however, circulation of the published volumes was not high. Unlike the other regions of AAA, the Midwest, because of its affiliation with MBAA, could not give copies to all registrants without asking for additional payment. For example the 1977 Proceedings cost $5, in addition to the $3 registration fee. The 1979 Proceedings cost $6 in addition to the $4 registration fee; the 1980 cost was $9, which was in addition to the $9 registration fee. Fewer than half of the registrants bought copies of the Proceedings (mostly just those who had papers included therein). In 1987, the

Proceedings of the Milwaukee meeting were given to all registrants because the independence that had been achieved from MBAA permitted the Region to charge a sufficient registration fee to cover the cost of the Proceedings.

Secession From the MBAA

The 1987 split from the MBAA was controversial in that many people liked the idea of holding the meeting regularly in Chicago because of its central location and the ease of making airline connections. In addition, the advantages of attending sessions sponsored by the other disciplines of MBAA were often mentioned. Alternatively, many people were sick of nine continuous years of the Windy City and the Palmer House. Those in the western and southern parts of the region were most anxious to secede from the MBAA because they felt the meeting should occasionally be held in St. Louis and Kansas City. There may also have been disagreement with the officer selection process of MBAA. Despite being the largest component of the MBAA, there had been only one accountant (Richard Metcalf of Drake University) who had served as MBAA president during the organization's history.

The subject of secession from the MBAA had been under discussion since at least as early as 1981. A July 8, 1981, letter from Joseph A. Silvoso (1980-81 president) to John H. Smith (University of Iowa), who had just been elected as the 1982-83 regional vice president, noted that although the Midwest Region was the largest of the AAA regions in terms of membership, it had the lowest percentage of membership attending the regional meeting. Silvoso speculated that the reason for the low participation was:

> that the membership generally does not regard the Midwest Regional Meeting of the Association as an accounting meeting. This situation exists primarily because the Midwest Regional annual meeting is part of the Midwest Business Administration annual meeting. When a region holds its meetings separately from other organization meetings as do all the regions with the exception of the Southwest and Midwest Regions, the proportion of the membership in attendance is high, there exists an espirit de corps, and the meeting is regarded as an important "accounting" meeting. With increasing travel costs, participation of our membership in a regional meeting will increase because an annual meeting of a region is more economically accessible generally than the annual meeting of the Association [Silvoso, 1981].
>
> Objections have been raised about the Midwest Region going its own way as Finance has done. One of the main objections is that the additional work will be required of the regional officers. This argument is false. In fact, the officers will have more time to plan and control provided they ask the Administrative Office of the Association to assist. I suggest you call Paul Gerhardt, the Association's Administrative Secretary,

and discuss with him the matter of assistance he and his staff can provide. Paul will help you in any way you desire.

So, John, I am suggesting to you to separate the Midwest Regional Meeting from the Midwest Regional Business Administration meeting. I believe you will be glad in 1982 when you chair the Midwest Regional meeting and circulate a brochure somewhat similar to the enclosure. We then can commence to truly have a regional accounting meeting.

Nothing immediately came of Silvoso's suggestion. The Midwest AAA continued to meet with the MBAA. In 1984, Paul Gerhardt wrote to Jack Krogstad encouraging him to support a split from MBAA. In 1985-86, under the administration of regional vice president Valdean Lembke (University of Iowa), the Midwest Region of AAA decided to split off from the umbrella organization. Lembke stated that there were basically two reasons for the split: the difficulty of coordinating the meetings through the MBAA, and the desire to get away from Chicago and the Palmer House. The meeting rooms at the Palmer House were, for the most part, too small to house the meetings of the accountants. The AAA was the largest of the MBAA constituent groups, but was often forced because of discipline politics into the smallest meeting rooms. In addition, Lembke felt that the miscellaneous charges at the Palmer House were ridiculously high, thus limiting alumni breakfasts and premeeting continuing education programs. Lembke urged the regional committee to secede from the MBAA in order for the AAA to have more control over its own program. There was no opposition to the proposal from the regional committee, but at the grassroots level there was a great deal of opposition. Lembke was prepared for a long and unruly business session at the 1986 meeting, and he so warned national president-elect Ray Sommerfeld, who was speaking at the business meeting. Sommerfeld stifled the opposition, however, by giving a speech on how the proper way to register complaints in a professional organization was to elect people to office who want to do things as you want them done. Standing up and creating a scene during the business meeting was not the democratic way of registering views. This speech so shocked the opposition that no one spoke out against the decision to leave the MBAA and the business meeting was one of the most tranquil on record [Lembke, 1990].

Interestingly, an accounting track has recently been added back to the MBAA. The name of the new MBAA accounting group is the Midwest Accounting Society (MAS). Paul Gerhardt notified the AAA Executive Committee of the new organization along with a probable explanation of why the MAS was formed. Gerhardt explained that a poll of exhibitors at the 1986 MBAA meeting indicated that 3/4 of the exhibitors were there because of accounting, and yet the exhibit revenue went to the MBAA, and not the AAA regional group. Gerhardt surmised that the newly formed MAS was an attempt to fill the void left in the MBAA meeting in hopes of avoiding a defection of exhibitors [Gerhardt, March 23, 1987].

Midwest Regional Vice Presidents

The regional vice president (or chairperson in the early years) was responsible for the development of the professional program through the 1983 meeting. Beginning in 1984, the vice-president-elect assumed the responsibility of program chairman.

Exhibit 4-5 lists all of the regional leaders and the locations of the annual meetings. Members of the region will notice that the list is not identical to the one that has appeared in the annual program booklet for the past several years. The program booklet has been in error since 1984 because the name of Louis Biagioni has been omitted. This naturally caused the dates to be wrong for several other early regional chairmen. The cause of that mistake was apparently a printing error, be-

Exhibit 4-5

Midwest Region Officers and Meeting Locations

Year	Location	Regional VP	Program Coor.
1960	Minneapolis, MN	Leon E. Hay, Indiana University	
1961	Indianapolis, IN	James M. Carrithers, Coe College	
1962	Chicago, IL	E. Joe DeMaris, University of Illinois	
1963	St. Louis, MO	Joe R. Fritzemeyer, State University of Iowa	
1964	Chicago, IL	James M. Fremgen, University of Notre Dame	
1965	Kansas City, MO	Louis F. Biagioni, State University of Iowa	
1966	Columbus, OH	Victor Tidwell, Kansas State University	
1967	Chicago, IL	Robert H. Raymond, University of Nebraska	
1968	Minneapolis, MN	Richard W. Metcalf, Drake University	
1969	Chicago, IL	Hadley P. Shaefer, Tulane University	
1970	Detroit, MI	Donald F. Istvan, Peat, Marwick, Mitchell & Co.	Thomas V. Hedges
1971	Chicago, IL	Glenn L. Johnson, University of Kansas	
1972	St. Louis, MO	Charles T. Andrews, Creighton University	
1973	Chicago, IL	William R. Welke, Western Michigan University	
1974	Chicago, IL	Richard Ortman, University of Nebraska at Omaha	
1975	Chicago, IL	John Sheppard, Western Michigan University	
1976	St. Louis, MO	Stephen L. Buzby, Indiana University	
1977	St. Louis, MO	Stephen W. Vasquez, St. Louis University	
1978	Chicago, IL	Harold Sollenberger, Michigan State University	
1979	Chicago, IL	Rick Elam, University of Missouri	
1980	Chicago, IL	Joseph Schultz, University of Illinois	
1981	Chicago, IL	Jack Gray, University of Minnesota	
1982	Chicago, IL	Kenneth O. Elvik, Iowa State University	
1983	Chicago, IL	John H. Smith, University of Iowa	
1984	Chicago, IL	Gale E. Newell, Western Michigan University	Jack L. Krogstad
1985	Chicago, IL	Jack L. Krogstad, Creighton University	Valdean C. Lembke
1986	Chicago, IL	Valdean C. Lembke, University of Iowa	Norlin G. Rueschhoff
1987	Milwaukee, WI	Norlin G. Rueschhoff, Notre Dame University	Lawrence C. Sundby
1988	St. Louis, MO	Lawrence C. Sundby, St. Cloud State University	Richard E. Baker
1989	Minneapolis, MN	Richard E. Baker, Northern Illinois University	Robert E. Holtfreter
1990	Chicago, IL	Robert E. Holtfreter, Fort Hayes State University	Paul M. Fischer
1991	Kansas City, MO	Paul M. Fischer, University of Wisconsin-Milwaukee	Thomas D. Hubbard

cause a copy of the original list, which was correct as submitted to the printer, was found in the Association files. The original program listing has then been copied year after year since then.

The early influence of Indiana University cannot be overlooked. Eight of the first 13 regional chairmen had an affiliation at one time or another with Indiana University. More recently, the center of influence has seemingly moved west as five of the past ten regional vice presidents have held doctorates from the University of Nebraska. The University of Illinois has surprisingly had little influence on the regional group as only five of the 32 regional leaders have had either a faculty or doctoral affiliation with the Champaign-Urbana school.

Unlike in some of the other regions, no Midwest Region vice president has yet gone on to become the AAA president. However, many have gone on to hold positions on the AAA Executive Committee. Interestingly, many of the regional leaders have left the region shortly after holding office. Apparently being regional vice president was a stepping stone to a better job. The first five regional leaders all left the region after being regional vice president. One regional leader, Hadley Shaefer (1968-69) left before finishing his term. Others, such as Charles Andrews (1971-72) left the same year their term ended. Altogether 11 of the first 13 regional chairmen left the region shortly after their term of office, and a 12th individual moved within the region. This mobility is somewhat unusual among AAA regional vice presidents. For instance, the older Southeast Region has had only one of its regional leaders leave the region—and that covers a 42-year period of time. It could be that the Midwest Region's slow growth in the early years is at least partly attributable to the loss of leadership from the region.

Midwest Region Summary

The Midwest Region of AAA has seen its annual meeting experience tremendous growth in recent years. The group outgrew the Midwest Business Administration Association and has since become quite successful as a separate entity. Much of the growth came by way of attendees from outside the region as nearly 1/3 of the speakers in recent years have been from a state outside of the Midwest. Some might argue that this intrusion of speakers from other regions is a detriment to those from within who would like to appear on the program. However, that argument is countered by the one that says the outsiders bring in new ideas and new opportunities for cooperation and growth. The outsiders have resulted in a much less provincial attitude at the regional meetings. Paul Gerhardt recently placed the "new" Midwest Region in perspective with the following lines:

> Attendance has been growing at the Midwest Regional Meeting since
> the separation and I believe the region is on the verge of becoming one

of the largest attended meetings, if not the largest in attendance, of all of the Association's regions. In addition, the Midwest Regional Group has done well financially since the separation [Gerhardt, April 5, 1990].

The leadership of the region has included some of the biggest names in academic accounting, but none has yet become national president. The only anomaly about the regional leaders is their mobility. Many have left the region after their term as a regional officer. Perhaps this mobility can be attributed to the pattern of population growth that has been experienced by the region in that the Midwest has not seen the population growth evident in the Southeast and Southwestern Regions. Because it is no longer affiliated with the MBAA, the Midwest AAA is on the threshold of major growth. It is the largest region and is now seeing the stability within its leadership that can set the stage for future dominance in the AAA structure.

THE WESTERN REGION

The present Western Region of AAA got its start when the AAA Executive Committee approved the formation of a Pacific Coast Regional Group at the August 1965 meeting. The motion for the formation of the region was tabled on the first day because the Association's Long-Range Planning Committee was studying the problem of regional groups and their geographical distribution. After an overnight adjournment, Walter Kell reported that the formation of the Pacific Coast Region would not be in conflict with anything that the Long-Range Planning Committee would recommend. Therefore, Charles Horngren moved for recognition of the Pacific Coast Regional group and passage was unanimous [Minutes, August 1965, p. 19]. The states included in the new region were California, Nevada, Utah, Arizona, Hawaii, Idaho, Oregon, Washington, and Alaska. Later, Colorado, Montana and Wyoming were added when the proposal for a Great Plains Region was abandoned for lack of interest.

The first meeting of the Pacific Coast Region was held on April 23, 1966. Warren C. Bray (California State College—Los Angeles) was chairman of the first regional committee. Other members of that first committee were Alan R. Cerf (Berkeley), Charles W. Lamden (San Diego State College), Oswald Nielsen (Stanford), Walter S. Palmer (University of Nevada-Reno), James W. Robertson (University of Santa Clara), and Lawrence L. Vance (Berkeley). Vance was at that time president-elect of the AAA. Technically, this was not the first west coast regional meeting; there had been an earlier regional meeting held in California. In 1943, the AAA cancelled its annual meeting because of wartime travel restrictions and hotel limitations. Instead, regional meetings were held in New Orleans, Cincinnati, Boston, Chicago, St. Louis, and Berkeley [Zeff, 1966, p. 64]. Thus, it could be said that the first Western Regional meeting, and those of some other regions as well, took place in 1943.

The fifth annual meeting in 1970 in Seattle was held in the Student Union building on the University of Washington campus. It was a single day program held on a Saturday in early May. According to the program brochure, the first speaker was William L. Campfield from the General Accounting Office. Campfield was AAA's first black national officer. AAA president Norton Bedford was the luncheon speaker. After lunch, there was a panel discussion entitled "Toward A New Introductory Accounting Course." The panel consisted of Gerhard Mueller (Washington), Vernon E. Odmark (San Diego State), and Robert T. Sprouse (Stanford).

By 1972 the much larger Western Regional meeting was attracting 450 registrants to Hayward, with an annual budget of over $4,000 (including $2,804 of registration fees). The remaining funds came from CPA firms ($600) and publishers [Johnson, Dec. 18, 1972]. The 1972 meeting was still a one-day session on a Saturday, but there was a Friday evening dinner for panelists. The opening plenary session featured keynote speaker Robert R. Sterling (University of Kansas). This was followed by a series of four concurrent sessions—all panel discussions. There was a luncheon address by Hector Anton (University of California, Berkeley).

In summary, Alan Johnson and the Hayward faculty made great strides toward expanding the offerings of the regional meeting. The 1972 activities were so well organized that the plans used by the Hayward faculty were incorporated in the AAA booklet published in 1974 entitled *Regional Meetings Planning Manual*.

By 1974, the meeting had expanded to a day-and-a-half duration, beginning with a luncheon on Friday. The format had also changed with the concurrent sessions consisting of presented papers rather than panel discussions. Based upon a list of participants that was prepared, there were at least 250 registrants attending the 1974 meeting. Also at the 1974 meeting an ad hoc committee was named to draw up a formal set of bylaws for the Western Region. Alan Johnson chaired the committee and prepared the preliminary draft of the bylaws. A final draft was sent to all members residing in the region and were presented for ratification at the 1975 meeting. The membership made one change to the draft before accepting the bylaws. Whereas the draft called the group the "West Region," the membership insisted (unanimously) upon retaining the title of "Western Region." [Fremgen, May 5, 1975].

The tenth annual meeting in 1975 was the second to be held at a hotel rather than on a university campus (the 1971 meeting in Las Vegas had been held at the Flamingo Hotel). All sessions in 1975 were held in the Holiday Inn on the beach overlooking Monterey Bay. The letter from James Fremgen to the membership announcing the activities of the meeting noted that there would not be a formal placement service because there had been minimal activity in previous years [February 24, 1975]. The letter also noted that the meeting would qualify for ten hours of continuing education credit. Fremgen's annual report of the Western Region's activities noted that there had been 207 registrants, 14 of whom were from community colleges. As a result of the low attendance from community colleges, Fremgen recommended that special sessions on that topic not be held in future years. Also, he noted that there had been difficulty in securing a speaker from a community college; no papers had been submitted and no community college faculty member would agree to speak [June 19, 1975]. Fremgen also suggested that it might be advisable to publish the proceedings of the meeting in future years.

Attendance dropped to 166 at the 1976 meeting in Tempe, Arizona. Apparently location was the cause of the decline in attendance because the following year's meeting in San Jose drew 300 registrants. The 1978 meeting in Pasadena attracted 323 registrants plus 27 spouses. The decade of the 1970s closed out with the meeting in Reno, where 228 participants registered. The meetings changed little during the late 1970s. There were no placement facilities offered, nor was there publication of the proceedings, although participants were instructed to make copies of their papers available at the meeting for those who were interested in a copy. The format of the 1978 meeting changed slightly by beginning earlier. A Thursday evening invitational dinner was held for program participants (hosted by Touche Ross and Co.) and the first program sessions began at 10:00 on Friday. The meeting then concluded with a luncheon on Saturday. A similar schedule was followed in 1979 in Reno. The program booklet for the Reno meeting was the most impressive one ever prepared for any AAA regional meeting. Whereas programs for all previous years had been published on a single sheet of paper, the 1979 program booklet, subsidized by the Sahara Reno Hotel, contained 44 pages—mostly in color. Actually, most of the contents was devoted to advertisements by firms in the Reno area.

There were slight changes to the bylaws at the 1979 meeting in Reno. Whereas the regional vice president had previously been from the university which hosted the annual meeting, the amended bylaws provided that the vice president could be from any school.

The 1982 meeting in Anaheim was preceded by a Thursday workshop presented by the AAA International Accounting Section. The international program was repeated in 1983, and was accompanied by two additional programs: "Deloitte, Haskins and Sells Auditscope Seminar" and "Teaching Not-for-Profit Accounting," offered by the AAA Public Sector Section (now Government and Nonprofit Section). The 1983 meeting included 3 plenary sessions plus 16 concurrent sessiions, each of which contained three papers. Altogether, 63 authors were represented on the papers presented. In just over a decade the Western Regional meeting had progressed from a meeting composed primarily of two or three panel discussions to a participative meeting involving a large number of members.

Revenues for the 1984 meeting totaled $19,461, which was in excess of expenses by $1,191. The 1985 meeting, with 248 registrants, generated revenues of $26,259; expenses were only $21,935, resulting in an excess of revenues over expenses of $4,324. Attendance at the 1986 meeting in Costa Mesa, CA was 346. Price Waterhouse financed the printing of the proceedings. The program booklet was a rather impressive 32 pages. The 1986 meeting generated revenues of $31,957, which resulted in an excess of revenues over expenditures of more than $10,000. By 1987, the Region was operating with revenues of over $48,000, about half of which came from registration fees and half from contributions and exhibitor fees. Expenditures were under $40,000. The Western Region had become a definite profit center.

The table below lists the locations of the regional meetings and the regional vice presidents (chairmen) and program coordinators. Surprisingly, none of the regional leaders has gone on to become a national president; however, several have joined the Executive Committee in other positions.

Exhibit 4-6

Year	Location	Regional VP	Program Coor.
1966	Berleley, CA	Warren C. Bray, Cal State-Los Angeles	
1967	San Francisco, CA	William Niven, San Francisco State College	
1968	Los Angeles, CA	A. N. Mosich, University of Southern California	
1969	Los Angeles, CA	John W. Buckley, UCLA	
1970	Seattle, WA	Thomas W. Porter, Jr., University of Washington	Committee
1971	Las Vegas, NV	Richard E. Strahlem, University of Nevada-Las Vegas	
1972	Hayward, CA	Alan P. Johnson, California State College at Hayward	
1973	Fullerton, CA	Robert W. Vanasset, California State Univ. at Fullerton	
1974	Sacramento, CA	Buquar A. Zaidi, California State at Sacramento	
1975	Monterey, CA	James M. Fremgen, Naval Postgraduate School	James M. Fremgen
1976	Tempe, AZ	Joe R. Fritzemeyer, Arizona State University	Charles Smith
1977	San Jose, CA	Joseph Mori, San Jose State University	Joseph Mori
1978	Pasadena, CA	E. Kennedy Cobb, California State University-LA	E. Kennedy Cobb
1979	Reno, NV	Benjamin L. Smith, University of Nevada-Reno	B. J. Fuller
1980	San Diego, CA	Kevin M. Lightner, San Diego State University	Kevin M. Lightner
1981	Fresno, CA	Elwyn L. Christensen, Cal State University-Fresno	Elwyn L. Christensen
1982	Anaheim, CA	Dale Bandy, Cal State University-Fullerton	Dale Bandy
1983	San Francisco, CA	Steven M. Mintz, San Francisco State University	Steven M. Mintz
1984	Tuscon, AZ	Russell M. Barefield, University of Arizona	Russell M. Barefield
1985	San Diego, CA	Wayne A. Label, University of San Diego	Wayne A. Label
1986	Costa Mesa, CA	Marc F. Massoud, Claremont McKenna College	Marc F. Massoud
1987	Sacramento, CA	Amin A. Elmallah, California State University	Amin A. Elmallah
1988	Monterey, CA	Kenneth J. Euske, Naval Postgraduate School	Shahid L. Ansari
1989	Las Vegas, NV	Ronald A. Milne, University of Nevada-Las Vegas	Ronald A. Milne
1990	Coeur d'Alene, ID	Jeffrey L. Harkins, University of Idaho	James Sepe
1991	Scottsdale, AZ	J. Hal Reneau, Arizona State University	Robert J. Capettini

Western Region Summary

The Western Region was somewhat slow to develop when compared to other AAA regions, but once it began operations, attendance and participation grew quickly. The great distances that many members must cover has held down attendance in comparison to total regional membership, but that cannot be avoided without splitting up the region. Another problem that the region has faced in recent years has been the lack of interest shown by professors at some of the larger schools in the region. Unlike other regions, professors at major doctoral granting institutions have not been altogether supportive of regional meeting activities. As a result, the program has been dominated by individuals from smaller schools. As an example, only ten of the papers at the 1988 meeting were authored by professors at doctoral granting schools within the region, and half of these people were from the University of Southern California. Also, there is a high percentage of participation from individuals outside of the region. At the 1988 meeting, 42 authors were from outside the Western Region. This participation from outside the region is to be expected given that the locations of most meetings are in places that are desirable vacation spots.

THE MID-ATLANTIC REGION

Although the Mid-Atlantic Region did not hold a formal meeting until 1975, the Region had been recommended as early as 1953 ["Report of the Sub-Committee, 1953, p. 326]. The Region formally existed on paper as early as 1966 under the name Middle Atlantic Region. The Association's Long-Range Planning Committee of 1966 presented a proposal for organizing the entire country into eight regional groups. There was also to be a Canadian group. The proposal was not altogether innovative since six of the eight regions were already holding regional meetings. Only the Middle Atlantic and the Great Plains regions were not already operative. The states that were included in the Middle Atlantic Region were Pennsylvania, New Jersey, Delaware, District of Columbia, Maryland, and West Virginia. This proposal was adopted by the Executive Committee at the May 1966 meeting. The report of the 1969 Committee on Regional Arrangements reiterated the need for a Middle Atlantic regional group and provided guidelines for a committee to be appointed by the president for the establishment of a meeting [Report of the Committee, 1969, p. 3]. However, due to the lack of leadership initiative, no meetings were held in the Middle Atlantic region for the next few years.

Finally, a leader came to the forefront in the person of Roger H. Hermanson, then of the University of Maryland. At the August 19, 1972 Executive Committee meeting, Robert Sprouse announced that he had taken action to organize the Great Plains and Middle Atlantic Regions. Apparently that action was to appoint Hermanson to chair a committee which would organize a Mid-Atlantic regional meeting. The Association provided a $600 budget allocation to subsidize a regional committee meeting. However, Hermanson wanted to spend the money for computer services to analyze the results of a questionnaire which dealt with the need for a regional meeting. Sprouse felt there was a definite need for a committee meeting and so authorized an additional expenditure not to exceed $600 for the meeting [Gerhardt, December 29, 1972]. The committee subsequently appointed included Hermanson as chairman, William Markell (University of Delaware), Anthony J. Mastro (George Washington University), Robert S. Maust (West Virginia University), G. Kenneth Nelson (Pennsylvania State University), and Charles J. Weiss (Seton Hall) [Hermanson, March 25, 1973].

The aforementioned questionnaire was sent to 1801 members of the Association residing in the states included in the region plus Virginia. The 778 responses consisted of 245 members in education, 233 in public practice, 143 in industry, and 129 in government (the low percentage of responses from educators, about 32%, was in reasonable proportion to the relative numbers of academic members at that time). Over 54% of the responses came from Pennsylvania and New Jersey, with an additional 18% from Maryland. Virginia, which ultimately was not included in the region, contributed 20% of the responses. The committee concluded that the responses supported the recommendation that the Middle Atlantic Region be activated. Attendance would be sufficient to justify this separate region without crippling the annual meetings of the Northeast and Southeast regions. The first meeting was planned for the spring of 1974. The committee report also included a recommendation that Virginia be included in the Middle Atlantic region [Hermanson, March 25, 1973].

On April 1, 1973, the Association's Executive Committee unanimously approved the activation of the Middle Atlantic Regional Group. The first meeting was to be held in 1974. According to the Minutes, there was then discussion regarding the region to which Virginia should belong. Robert Anthony moved as follows:

> Virginia shall officially be in the Southeast Regional Group but it will also be included in all mailings for the Middle Atlantic Region.

James Don Edwards seconded the motion and it passed unanimously.

Unfortunately, due to a number of circumstances, the first meeting was not held in 1974. First, in the summer of 1973, Roger Hermanson, the catalyst for the new organization, moved to Georgia State University, which was outside the region. Next, a letter in the Association's archives shows that President Robert Anthony appointed Stephen E. Loeb of the University of Maryland as the first regional vice president of the region. The letter begins with a statement that "Roger Hermanson has told me of the great assistance you have been to his Committee in making the study that led to the creation of the Great Atlantic Region of the American Accounting Association" [Anthony, July 3, 1973]. It is uncertain from where the phrase "Great Atlantic Region" originated. However, Loeb declined to serve. Anthony then asked William Markell of the University of Delaware who agreed to serve if the first regional meeting could be postponed until 1975 [Markell, July 23, 1975].

The Association's archives are subsequently silent with regard to the matter until April 19, 1974, when a Mid-Atlantic Regional Steering Committee meeting was held in Philadelphia (during the Northeast Regional meeting which was held in that city). Those in attendance at this first organizational meeting were:

R. Lee Brummet, President-elect, AAA
Stephen Loeb, University of Maryland
William Markell, University of Delaware (Chairman)
Anthony Mastro, George Washington University
Bart Basi (representing Ken Nelson), Pennsylvania State University
Bruce Oliver, University of Pennsylvania
Jack Topiol, Community College of Philadelphia

Apparently the University of Delaware had already been decided upon by this time since the first item in the Minutes was to suggest that Robert Paretta be appointed to the Steering Committee to act as secretary. It had been decided that the secretary should be someone from the host school who would assist the regional vice president.

Most of the committee's time was spent discussing the 1975 meeting. Of foremost importance was to select a date that would not conflict with the meetings of either the Northeast or Southeast meetings. It was decided that the meeting should be of one-and one-half days duration similar to that arranged by the Northeast Regional Group.

Steve Loeb suggested that as something innovative and unique, the Region should consider preparing a Proceedings of the meeting and have it ready for members as they arrived. This would require an earlier call for papers than had been traditional for other regional meetings. Loeb volunteered the University of Maryland to handle the details, including binding. Just the idea to publish a proceedings was an innovative idea at the time because no other region had yet published a proceedings (other than the Canadian Region, and this fact was not known in the States). The Southeast Region published its first proceedings following the April, 1974 Memphis meeting. The Memphis volume was mailed to registrants after the meeting.

The 1975 meeting proved to be a success. As planned, the Proceedings, edited by Robert Paretta, were distributed at the beginning of the meeting; this was the first time that a region was able to provide registrants with a copy of the Proceedings in advance of the sessions. The Proceedings were financed by a contribution from the firm of Price Waterhouse. The program consisted entirely of concurrent sessions, with the exception of a luncheon speech by Abraham Briloff. Briloff's address was later cited as being the highlight of the meeting [Markell Letter, July 23, 1975]. The Association was represented by Yuji Ijiri who attended at the request of president-elect Wilton Anderson. There were 182 registrants for the meeting. Of these, 96 names appeared on the program including discussants and session chairpersons. There were a total of 17 different sessions throughout the program with three concurrent sessions being held at a time (with the exception of the last hour on Saturday morning).

The 12 sessions devoted to the presentation of papers included a total of 24 papers. These were heavily devoted to financial accounting and accounting education. There were no papers in the areas of taxation, accounting systems, accounting history, international accounting, or governmental and non-profit accounting—all subjects that are now staples at the regional meetings. Governmental accounting is probably the most obvious exclusion from the program since the Region includes Washington, D. C. and the original Steering Committee wanted to include governmental accounting as a major part of the program.

The 1975 meeting proved to be a financial success (as were all future meetings except two). A profit was generated in the amount of $1,637.80, which was passed along to the 1976 regional vice president, G. Kenneth Nelson of Pennsylvania State University.

Subsequent Meetings

Meetings in subsequent years were in many respects similar to that of 1975. Prior to 1989, attendance varied from 160 to 225 registrants. The 1989 attendance was a record at 334. The Proceedings continued to be distributed at the meeting, and continued to be financed by Price Waterhouse.

The 1975 program booklet was quite large and was printed on a total of three sheets of paper. In subsequent years, however, the program listings were consolidated onto one sheet of paper. That tradition was broken in 1984 when the program booklet for the University of Baltimore meeting was enlarged to an impressive 60 pages. Included were welcomes from the mayor of Baltimore and two college presidents. There were also book advertisements and lists of presenters and moderators. However, the program was the poorest in some respects in that it did not provide the name of the regional vice president or the names of the members of the steering committee for the meeting. All prior programs had included such information. The 1985 program reverted to the older, smaller style, although it was printed on three sheets of paper.

Another highlight in the region's history came in 1987. At the October 31, 1986, Executive Committee meeting of the Association, a resolution was passed requiring the various regions to transfer all regional funds to the national office. The funds were to be maintained in the Association's bank account and disbursed upon authorization from the regional officers. This motion resulted from a recommendation from the Association's auditors who expressed concerns regarding the accountability over regional group funds maintained at many locations throughout the country. All of the regions were quick to comply with this directive—with the exception of the Mid-Atlantic. The 1987 Mid-Atlantic steering committee voted to defy the directive and keep $5,000 in the hands of the regional treasurer for "working capital." The Association's Executive Committee expressed concern over the problem at the November 6, 1987, meeting, but felt that no action was necessary because the Mid-Atlantic Region would likely fall into line during 1988. Following the 1988 regional meeting in University Park, PA, J. Edward Ketz transferred $1,450 to the national office to keep the amount transferred to the next program director at "about $5,000." Finally, on October 20, 1988, Araya Debessay transferred the remaining regional balance of $5,080.88 to national in order to be able to report that the Mid-Atlantic Region was in compliance with AAA policy.

Mid-Atlantic Regional Officers

The Mid-Atlantic Region has maintained a tradition of having the regional vice president and the program coordinator come from the host school. Exhibit 4-7 lists the host schools, vice presidents, and program coordinators over the years since

Exhibit 4-7
Mid-Atlantic Region Officers and Locations

Year	Location	Host	Regional VP	Program Coor.	Regis.	Ret. Earnings
1975	Newark, DE	University of Delaware	William Markell	Robert L. Paretta	182	$1637.80
1976	University Park, PA	Penn State University	G. Kenneth Nelson	Lawrence Klein	210	$2365.30
1977	Washington, DC	George Washington Univ.	Anthony J. Mastro	Michael Gallagher	160	$2840.38
1978	South Orange, NJ	Seton Hall University	Charles J. Weiss	Jeremiah G. Ford	206	$4108.22
1979	College Park, MD	University of Maryland	Stephen E. Loeb	Lawrence Gramling	225	$7585.38
1980	Morgantown, WV	West Virginia University	Robert S. Maust	Horace Givens	187	$6936.97
1981	Clarion, PA	Clarion State College	Charles J. Pineno	S. Theodore Hong	225	$7538.24
1982	Secaucus, NJ	Fairleigh Dickinson Univ.	John Cerepak	Philip Blanchard		$7924.64
1983	Philadelphia, PA	Drexel University	Cadambi Srinivasan	H. Rao/J. Savchak	217	$8279.15
1984	Baltimore, MD	Un. Balt./Catonsville Comm. College	Arthur T. Roberts	Manuel A. Tipgos		$7442.63
1985	Washington, DC	George Washington Univ.	Michael Gallagher	Anthony J. Mastro	183	$7737.23
1986	New Brunswick, NJ	Rutgers University	Bikki Jaggi	Yaw M. Mensah	170	$11,722.75
1987	Atlantic City, NJ	Temple University	Stephen L. Fogg	Bill Schwartz		$23,444.56
1988	University Park. PA	Penn State University	J. Edward Ketz	J. Edward Ketz	176	$25,630.73
1989	Wilmington, DE	University of Deleware	Araya Debessay	Araya Debessay	334	
1990	Washington, DC	Howard University	Phillip Blanchard	Margaret Hicks		
1991	East Hanover, NJ	Fairleigh Dickinson Univ.	John Cerepak	Robert A. DeFilippis	170	

the region became active. Also listed are the official attendance figures for each year and the amount of funds transferred annually to the next year's vice president. The attendance figures for some years were not available apparently due to the vice president failing to submit the annual report (there were copies in the Association's files of second-request letters mailed to the vice presidents).

The officers who organized the 1981 meeting at Clarion State College were Charles J. Pineno and S. Theodore Hong. However, these individuals were not listed as the officers who were elected in 1980. At the election in 1980, Curtis E. Bagley of Clarion State was elected vice president with Forest C. Carter as the program coordinator. However, Bagley left academe to go into industry. He attended the 1981 meeting as a representative of Corporate Jet, Inc.

Another unique aspect of the list of officers occurred in 1988 and 1989 when both J. Edward Ketz (Pennsylvania State University) and Araya Debessay (University of Delaware) fulfilled the duties of both the regional vice president and the program coordinator. The work of the two positions was again divided for the 1990 meeting.

Summary of Mid-Atlantic Region

The Mid-Atlantic Region was slow in starting, and has seen its attendance figures waver somewhat from year to year. Despite this, the regional meeting has grown in size (the 1989 meeting attracted 334 registrants resulting in many people not receiving copies of the Proceedings). The meeting has been profitable in every year but two (1980 in Morgantown, WV and 1984 in Baltimore). The regional meeting has traditionally boasted good programs and considerable participation from individuals from outside the region. In one respect, the meeting is different from those in other regions in that it does not have a large and active placement service. Whereas other regions have to set aside special rooms for the placement service, the Mid-Atlantic has only a file of a few openings placed on the registration desk. To many people, this is a disadvantage of the meeting; to others, it is viewed as a favorable situation.

Although still young, the Mid-Atlantic Region is an integral part of the American Accounting Association. Those individuals such as Roger Hermanson, William Markell, and the other members of the initial steering committee are to be applauded for the initiative they took in the early 1970's to activate the region.

SUMMARY AND CONCLUSION

Besides the seven regions now existing, there was also a Canadian region for a few years. The history of that group is included in the chapter on International

Outreach programs. There have also been several other proposed regional groups, but none have materialized. The Long Range Planning Committee of 1966 recommended the establishment of a Great Plains regional group, but no one in that region took the initiative. Thus, the Executive Committee, in March 1973, assigned the states of the Great Plains to the Midwest and Western regions. Interestingly, an August 1973 AAA committee report still urged the establishment of a Great Plains group to include the states of Montana, North Dakota, South Dakota, Wyoming, Nebraska, Missouri, Kansas, and Colorado [Minutes, August, 1973]. A practitioner member from Missouri tried to establish such a region in 1974, but was told by Paul Gerhardt that it was too late; the Executive Committee had already ruled out the idea.

In 1971, Don DeCoster and Gerhard Mueller of the University of Washington, tried to assess the interest in a Northwest Region (independent of California, Nevada, and Arizona). A contemporaneous Long Range Planning Committee recommended such a region be formed to include Washington, Oregon, Idaho, and Alaska [DeCoster and Mueller, 1971]. Interest in such a group was apparently lacking, however, as the group never materialized.

A British region was recommended in 1981 by L. G. Tostevin of Kensington University [Tostevin, 1981]. Paul Gerhardt nixed that idea by responding to Professor Tostevin that there were only 186 members in Great Britain, and that this number was insufficient to support a Section. His thinking may have been influenced by the fact that the Canadian Region, with more than 300 members, was at that time having major problems and was about to be abolished.

For many members of the AAA, the regional meetings are at the heart of the Association's activities. The regional meetings have grown ever larger in recent years, probably because of the high level of member participation permitted at the meetings. Each of the regions has a rich history, and perhaps surprisingly, there is little similarity among the histories of the various regions. Each region is distinctive, and has its own distinct history.

References

"AAA Committees," *The Accounting Review*, January, 1954-1973.

"AAA Regional Meetings Gain Interest," *The Accounting Review*, April, 1960, pp. 363-366.

American Accounting Association Program, Midwest Regional Meeting, 1988, and 1989.

Buzby, Stephen L., *Collected Papers of the 1976 Annual Meeting Midwest Region American Accounting Association*, St. Louis, 1976.

Committe, Bruce, Hassan Espahbodi, and Gary John Previts, *A History of the Southeastern Regional Group of the AAA*, (Sarasota, FL: American Accounting Association, 1978).

DeCoster, Don, and Gerhard Mueller, Letter to Fellow Accountants, November, 15, 1971.

Epaves, Richard A. and Joseph M. McKeon, Jr., *A History of the Ohio Region of the American Accounting Association*, Oxford, Ohio: American Accounting Association Ohio Region, 1984.

Fremgen, James M., Letter to R. Lee Brummet, May 5, 1975.

Fremgen, James M., Letter to Members of the Western Region, February 24, 1975.

Fremgen, James M., Report to Executive Committee, June 19, 1975.

Gerhardt, Paul L., Letter to David Solomons, November 3, 1969.

Gerhardt, Paul L., Letter to Roger H. Hermanson, December 29, 1972.

Gerhardt, Paul L., Letter to Executive Committee, March 23, 1987.

Gerhardt, Paul L., Letter to Larry Kreiser, April 5, 1990.

Hay, Leon, Letter to Dale L. Flesher, June 14, 1990.

Hermanson, Roger H., Letter to Paul Gerhardt, March 25, 1973.

Hood, James, Interview by Dale L. Flesher, October 24, 1990.

Johnson, Alan P., Letter to Robert Sprouse, December 18, 1972.

Lembke, Valdean, Interview by Dale L. Flesher, September 7, 1990.

Markell, William, Letter to R. Lee Brummet, July 23, 1975.

Midwest Business Administration Association Program, Various Years.

"Midwestern Group Meeting," *The Accounting Review*, October, 1960, p. 758.

"Minutes of the Executive Committee Meetings," (Sarasota, FL: American Accounting Association, 1966-89).

Most, Kenneth, Interviewed by Dale L. Flesher, April 1, 1990.

1989-90 Directory of the American Accounting Association (Sarasota, FL: American Accounting Association, 1989).

"1987 Midwest Regional Group Meeting," *Accounting Education News*, October, 1986, p. 18.

"1987 Mid-West Regional Group Meeting," *Accounting Education News*, June, 1977, p. 9.

Powell, Ray M., *Program Planning Manual for AAA Regional Meetings* (Sarasota, FL: American Accounting Association, 1974).

Previts, Gary John, "Some Relevant Factors About the History of the Southeastern Regional Group," *Collected Papers of the Twenty-ninth Annual Meeting*, (Nashville: Southeast Regional Group American Accounting Association, 1977), pp. 20-24.

"Report of the Committee on Regional Arrangements," (Sarasota, FL: American Accounting Association, 1969).

"Report of the Sub-Committee of the AAA Executive Committee on Regional Meetings," *The Accounting Review*, July, 1953, pp. 325-326.

Shillinglaw, Gordon, Letter to Dale L. Flesher, June 14, 1990.

Silvoso, Joseph A., "Letter to John H. Smith, July 8, 1981.

"Southwestern Group Meeting," *The Accounting Review*, July 1961, pp. 511-512.

"Southwestern Section," *The Accounting Review,* October, 1960, pp. 756-757.

Sprouse, Robert T., "Comments From the President," *The Accounting Review,* January, 1973, p. 175.

Starling, Jack M. and Joan D. Bruno, *Southwestern Federation of Administrative Disciplines: Statistical Report and History,* Dallas: SWFAD, 1990.

Tostevin, L. G., Letter to Paul Gerhardt, April 1, 1981.

Wood, Porter, "The Northeast Regional Group: A Brief History," *The Accounting Journal,* Vol. 1, No. 1, 1977, pp. 64-66.

CHAPTER 5
SPECIAL INTEREST SECTIONS

One of the more controversial subjects during the recent history of the AAA was whether special interest sections should be allowed to form. There was fear among members of the various Executive Committees that the formation of special interest groups would result in a splintering of the organization. The idea of sections has emerged several times over the years. For example, Myron Gordon suggested at a 1965 meeting of the Executive Committee that management accountants needed a special interest section because they received little benefit from being members of AAA since there were so few managerial accounting articles published in *The Accounting Review*.

An early attempt at doing something about the special interest problem was implemented by James Don Edwards when he allotted time for special interest groups on the afternoon preceding the 1971 annual meeting in Lexington ["Comments...," 1971, p. 390]. Subsequent presidents offered similar opportunities. Also, President Robert Sprouse appointed special interest committees which studied problems in specialized areas such as taxation, managerial accounting, and accounting history.

Ultimately, it was Gary J. Previts (University of Alabama), S. Paul Garner (University of Alabama), and Alfred R. Roberts (University of Missouri) who were responsible, albeit indirectly, for the AAA moving to an acceptance of sections. It was Previts, Garner, and Roberts, along with five others on the initial chartering committee, who established the Academy of Accounting Historians in 1973. Part of the motivation for establishing a new organization for accounting historians was the fact that the AAA Executive Committee had ignored various committee reports over the years (since 1968) which recommended more Association involvement in accounting history. Since the AAA was not doing anything, it was felt by the organizers of the Academy that a new organization sensitive to the issues of accounting history was needed [Coffman, 1989].

The splintering off of the historians might not have been so alarming to the Executive Committee had not D. Larry Crumbley, then at the University of Florida, noted the ease with which the Academy had been formed. Crumbley had been voicing displeasure over the fact that tax professors were not having a large enough voice in the AAA and were not given sufficient time on the annual program. Crumbley had been chairman of the Association's Federal Taxation Committee during the 1972-73 year under President Sprouse and felt the committee did not get enough attention from the Executive Committee. Then, Sprouse's successor, Robert Anthony, failed to appoint such a committee. At that point Crumbley became inspired by the Academy activities of Previts and Roberts. Crumbley copied the Academy's by-laws and used them to incorporate the American Taxation Association (ATA) in 1974 [Crumbley, 1989, p. 22].

Crumbley's organizational letter sent to tax professors emphasized the inequitable treatment given tax people by the AAA. One sentence stated "now is the time for the orphan of the accounting profession to seek tax power." The letter may have been intentionally emotive in terminology, but it did express the feelings shared by many tax professors [Crumbley, 1987, p. 87]. As a result, the ATA grew quickly, as did the Academy of Accounting Historians. Fearing that the AAA would lose these members, and perhaps others in specialized disciplines, the Executive Committee finally addressed the issue of sections.

Actually the first attempt at establishing a separate section involved a Young Professors Section which was recommended at the December 1973 Executive Committee meeting. It was noted that many young professors were already involved in the AAA and there was no need for a separate section, thus no action was taken [Minutes, December, 1973, p. 23].

Sectional matters began to take center stage at the Executive Committee meeting held August 15 and 16, 1974, when secretary-treasurer Robert Sweeney (University of Alabama) distributed a policy statement on sections and recommended adoption. His motion, which was seconded by Doyle Williams (Texas Tech), contained the following provisions:

1. Where sufficient interest has been demonstrated, the Executive Committee may authorize the development of interest groups or sections within the Association.
2. Each group would be required to submit to the Executive Committee a statement of its objectives and purposes which must be consistent with those maintained in the by-laws of the Association.
3. Each group will meet at the time of the Annual Meeting of the Association. The President of the Association may invite a group to conduct a concurrent session at the Annual Meeting.
4. Groups will be encouraged to work with the regional vice presidents relative to regional meetings of the Association.
5. The academic vice president will provide liaison between the groups and the Executive Committee.
6. The objectives, purpose and activities of a group will be considered in determining the financial support provided by the Association.

The discussion of the proposal addressed a variety of points. Robert Anthony thought that the sectional topics offered on the Monday afternoon of the annual meeting was enough of an effort toward special interest groups. Several members suggested that sections might fracture the Association. Also, several members did not want to give any financial support to entities which would be separate from the Association. The vote on the motion was three in favor, three opposed, and three abstentions. President Robert Anthony then cast a vote in favor of the motion and it passed. James Don Edwards then urged reconsideration of the motion. He stated that an issue as important as this should not be decided by such a split vote. He felt the matter should be considered further. Anthony then stated that the idea of sections could be called to the attention of members at the Monday afternoon "rap session" to be held at the annual meeting to get the views of members. Edwards moved that the motion be reconsidered; Robert Sterling seconded the motion and it passed unanimously. A revote was then taken on the previously passed motion and it was unanimously defeated. Sweeney then moved that the president-elect and the 1974-75 Executive Committee be heartily encouraged to investigate the matter of sections further. That motion was seconded by R. Lee Brummet and passed unanimously [Minutes, August 1974, pp. 15-17].

The same six-point proposal that had been made at the August meeting was offered again at the December 1974 meeting. In support of the proposal, Doyle Williams argued that the creation of sections would be a good way of enhancing membership service. Alternatively, Charles Zlatkovich and K. Fred Skousen argued that it was their opinion that the membership was not in favor of sections. Skousen would rather have spent the resources on the regions. Zlatkovich suggested that more input was needed from the membership before making a decision regarding the establishment of a sectional form of organization. Ultimately, when the motion was voted upon, the result was four in favor, four opposed, and one abstention. President R. Lee Brummet declined to cast a tie breaking vote, resulting in the failure of the motion [Minutes, November, 1974, pp. 10-11]. Brummet later stated that he was in favor of the motion, but feared that if it passed with less than a majority of the Executive Committee in favor, the concept would never be able to achieve its potential. Thus, he declined to break the tie in order to allow more time to build up support for the idea of sections [Brummet, 1989].

Following the failure of the motion, Zlatkovich moved that the December issue of *Accounting Education News* include a description and explanation of special interest sections and a ballot to be used to obtain a response from members regarding their preference on whether sections were desirable [Minutes, December, 1974, p. 11]. An article entitled "Sections: Pro or Con?" did appear on the front page of the December issue. It was noted that the history and tax groups had already been formed outside of the AAA structure and that there was "clearly some sentiment in favor of a move to sections." The advantages of sections were stated as follows:

> A section organization would provide a basis for identification and closer interaction between those of our members of common special interests in the field of accounting. It would provide a vehicle for delegation of certain education and research activities and responsibilities of our Association and perhaps for program planning of our regional meetings and our Association's annual meeting.

The disadvantages were given equal time with this statement:

> It is possible, on the other hand, that proliferation into sections could be divisive by detracting from our strength that comes from a more monolithic unity. This may be particularly relevant in view of the fact that a majority of our members are not professors and that only a minority of our members are active participants in our committee work and our annual meetings.

Readers were asked to mark the enclosed ballot which asked if the members were in favor of the AAA encouraging the development of sections, whether the individual would participate in sections if they were organized, and the specific sections (not to exceed three) that they would be most interested in joining ["Sections...," 1974, p. 1].

Paul Gerhardt reported at the March 1975 Executive Committee meeting that there had been 573 ballots returned. Of these, 412 replies were in favor of encouraging the development of sections and 161 replies were opposed. The same six-point proposal that had been presented at the two preceding meetings was again made. The vote was four in favor, two opposed, and one abstention. Interestingly, Robert Sweeney, who had made the proposal at the August meeting, was absent from the March meeting [Minutes, March, 1975, pp 10-11]. Thus, the proposal passed without one of its biggest supporters. The April newsletter contained a front-page story giving the results of the mail ballot and announcing the action of the Executive Committee ["Sections," 1975, p. 1].

Formal guidelines pertaining to the establishment of sections were developed by Paul Gerhardt and approved at the August 1975 meeting of the Executive Committee. Some of the provisions included in the first "AAA Section Policy" included a provision that a section's dues could not exceed more than one-half of the amount of the AAA national membership dues and a requirement that a section had to have 100 interested members before it could be established. A section could be discontinued at any time if its membership fell below 50 members [Minutes, August 14-15, 1975, Appendix C].

At the March 1978 Executive Committee meeting, a Committee on Regions and Sections recommended several amendments to the Association's policies with respect to sections. One provision was that fund raising had to be coordinated by the Executive Committee. Subsequently, at the March 1980 meeting, Stephen Zeff proposed that a *de minimus* rule be established which would allow each section, region, and group to solicit and accept contributions up to $2,500 per year without AAA Executive Committee action. That motion passed unanimously.

The formation of sections was an event waiting to happen. Individuals throughout the USA quickly began securing the necessary 100 signatures to form a variety of sections. Over the next few months members were notified of the sections being started through announcements in *Accounting Education News*. Subsequently, at the 1976 annual meeting in Atlanta, six sections held their organizational meetings. These first sections were the Auditing Section, the International Accounting Section, the Public Sector Section (now Government and Nonprofit Section), the Management Advisory Services Section (now Information Systems/ Management Advisory Services), the Community/Junior College Section (now Two-Year College Section), and the Administrators of Accounting Programs Group ["Sections Organize," 1976, pp. 1-2]. In later years, other sections were formed, including the Accounting, Behavior and Organizations Section, the Management Accounting Section, the Public Interest Section, and the Gender Section. In addition, the American Taxation Association returned to the fold as a section in 1978.

Despite the fears of certain members of the Executive Committee, the establishment of sections has not splintered the Association. The AAA is stronger than ever, probably because of the many membership benefits offered by the sections. Indeed, it is possible that a failure to establish sections could have been detrimental to the organization in that the sections would have been formed anyway, but outside the AAA structure (as did occur with accounting history and taxation). Ultimately, the movement toward sections led to a change in the governing structure of the Association in that the Council was formed in 1978 to permit the sections and regions to have a voice in the activities of the organization. In summary, what was feared would cause a break-up of the Association has led to a greater democratization of the group. Members formerly felt estranged from the inner workings of AAA, but with sections, the members are closer to the Executive Committee than ever before.

The true picture of how successful the sections have been can be seen by viewing the history of each of these groups. The following pages summarize the highlights of each section's founding and development. All of the sections are covered in this chapter with the exception of the International Accounting Section which is discussed in Chapter 2 dealing with the AAA's international initiatives. It should be noted that the present list of sections is likely not going to be complete for long. There are campaigns underway to start other sections including one for individuals interested in expert systems, another specializing in accounting education, and one in the area of financial accounting. Others are likely to follow.

ADMINISTRATORS OF ACCOUNTING PROGRAMS GROUP

One of the first group of special interest sections formed in 1976 was the Administrators of Accounting Programs Group (AAPG). A committee began working on the organization of the AAPG at the 1975 AAA annual meeting in Tucson. That committee was composed of Vincent Brenner (Louisiana State University), Paul E. Dascher (Drexel University), Joe R. Fritzemeyer (Arizona State University), William Markell (University of Delaware), Arthur G. Mehl (Bradley University), and Doyle Z. Williams (Texas Tech University). Arthur Mehl was chairman of the committee. The group was approved by the AAA Executive Committee at its March 1976 meeting, but only after considerable discussion of the merits of the proposed organization. Some members of the Executive Committee did not feel that the subject matter of AAPG was suitable for sectional status. As a result, the AAPG was approved by a 5-to-4 vote.

The organizational meeting of the new group, which was attended by about 150 individuals,[1] was held on August 23, 1976, in Atlanta at which time the bylaws were adopted and the first officers were elected [Mehl, June 29, 1976]. Four members of the organizing committee were elected to serve as the first officers, with Mehl as the first president. The officer positions originally consisted of a president, vice-president, treasurer, and secretary; a vice president of accreditation was added in 1988. Five individuals were elected to the initial board of governors, one of whom was Joseph A. Silvoso who was elected to fill the position of the past president on the board. Thus, Silvoso has never been president of the AAPG, but he has served as past president. A complete list of all AAPG national officers is shown in Exhibit 5-1. The positions of regional chairmen were added in 1981; that title was changed to regional vice president in 1985.

Unlike the other special interest sections of AAA, the AAPG is not open to everyone. In fact, there initially could be only one individual member from each institution.[2] Because of its limited membership potential, funding would have been a problem due to the AAA's requirement that annual dues not exceed 50% of the Association dues. Thus, at the time it was formed, the AAPG received special exemption for the limitation on dues [Anderson, April 6, 1976]. For example, the 1990 dues were $50—an amount greater than the Association dues of $45. Because of the restricted

---

[1] Many of these interested persons must have been discouraged by the $50 dues, because by April 1977 there were only 119 dues-paying members.
[2] A 1985 bylaws change established a professional membership status to enable representatives of accounting firms and other organizations to participate in AAPG activities. In 1990, another change was made to the bylaws which permits former chairmen to participate in AAPG. However, the membership is still institutional, only one dues payment is made, and each institution still has only one vote.

membership and the waiving of the limit on dues, the organization is not called a "section" within the AAA structure, but a "group." However, there is no real significance to the difference in terms other than the exemption from certain sectional guidelines. The only time that the term "group" has presented a problem was when the AAA Advisory Council was established in 1978. The motion which established the Council stipulated that the membership would consist of "regional vice presidents and section chairmen (for the upcoming year) or their delegates" [Minutes, March, 1978, p. 9]. Since the AAPG was not a section, no representation was allowed on the first Council. The omission of the AAPG from the Council was not a mere oversight. According to the March 1979 Executive Committee minutes, AAPG "was not intended to be included in the Advisory Council since it was a group and not a section." William Markell, 1979-80 AAPG president, campaigned for the AAPG to have a vote on Council, and such status was attained in November 1979.

The objectives of the Group are to promote the continuing development of high quality accounting programs; to provide a forum for the exchange of ideas; to facilitate and participate in the accounting accreditation process; and to facilitate mutual assistance of administrators of accounting programs. The official activities of the AAPG are to include:

a. Sponsoring educational programs.

b. Promoting research relating to the administration of accounting programs.

c. Publishing materials of interest to administrators of accounting programs.

d. Issuing Position Statements in the name of the Group on matters which impact on accounting education and its administration.

e. Active participation in the accreditation of accounting programs.

Exhibit 5-1
Administrators of Accounting Programs Group (AAPG) Officers

Year	President	Vice President	Secretary	Treasurer	News Editor
1977	Arthur G. Mehl	Paul E. Dascher	William Markell	Joe R. Fritzemeyer	Vincent J. Brenner
1978	Doyle Z. Williams	K. Fred Skousen	Henry R. Anderson	Jack E. Kiger	Charles J. Weiss
1979	K. Fred Skousen	William Markell	Stephen E. Loeb	Robert Williamson	Spencer J. Martin
1980	William Markell	Clarence Avery	Charles J. Weiss	Joseph E. Mori	Belverd E. Needles
1981	Clarence Avery	Joseph E. Mori	Charles Carpenter	Belverd E. Needles	Harold M. Nix
1982	Joseph E. Mori	Charles Carpenter	Gary E. White	Gary A. Luoma	Richard Samuelson
1983	Charles Carpenter	Gary E. White	James T. Hood	Harold Sollenberger	William C. Boynton
1984	Gary E. White	Harold E. Wyman	Maurice Stark	Patrick R. Delaney	James R. Davis
1985	Harold E. Wyman	Russell J. Petersen	James R. Davis	Jacob B. Paperman	L. Todd Johnson
1986	Russell J. Petersen	Robert E. Schlosser	Frederick Neumann	William F. Bentz	James R. Davis
1987	Robert E. Schlosser	Melvin C. O'Connor	Jan Williams	Keith Bryant	Gary W. Heesacker
1988	Melvin C. O'Connor	William F. Bentz	Paul A. Janell	Donald J. Klein	Gary W. Heesacker
1989	William F. Bentz	C. Dwayne Dowell	Bruce Oliver	Daniel O'Mara	Leonard E. Berry
1990	C. Dwayne Dowell	Jan R. Williams	James E. Smith	Lanny G. Chasteen	Jerry E. Trapnell
1991	Jan R. Williams	James E. Smith	Michael A. Diamond	Penelope J. Yunker	Jerry E. Trapnell

The first four of the above were included in the 1976 by-laws. The provision regarding accreditation was officially added in 1985, although it had been a major activity from the beginning.

Seminars

The Group has not been particularly active in some areas, but the sponsoring of educational programs has been quite productive. Under the sponsorship of the Continuing Education Committee, the annual winter seminar for administrators of accounting programs, which began in February of 1979, has become quite popular. Of course, the seminar is usually held in February in a warm climate, a factor that might help attendance. The first seminar in 1979 attracted 120 attendees (total membership was at that time 221). In recent years, over 200 have been in attendance (out of a total 1990 membership of 291). Two identical seminars were held in 1980—one in Orlando and one in Las Vegas—and in subsequent years through 1984. The annual offering was cut back to one program per year in 1985 with annual shifts between the west coast and Florida. A special program for new administrators was added in 1984 on the day preceding the start of the regular program. This pre-seminar half-day program has grown into a popular feature, perhaps due to the continual turnover among accountancy program administrators.

In addition to the national seminars, there are also seven regional seminars each year—one at each of the seven regional meetings. Total attendance in 1990 at these seven meetings was 210 individuals—just slightly more than the attendance at the February seminar in Florida. The regional meetings are planned and organized by the regional vice presidents.

Accreditation

Several committees were established at the first board meeting in 1976, including:

Committee on Accreditation and Standards of Professional Schools and/or Programs in Accounting (chaired by Catherine Miles, Georgia State University),

Committee on Research and Publication (chaired by Vincent Brenner, Louisiana State University),

Committee on Organization and By-Laws (chaired by John Cerepak, Fairleigh Dickinson University),

Committee on Continuing Education (chaired by Henry Anderson, Cal State—Fullerton).

The first of the above committees was the most important because the AICPA was considering taking over the accreditation process for accounting. At the time, the American Assembly of Collegiate Schools of Business (AACSB) was opposing autonomous schools of accounting, separate accounting accreditation, and separate control of the accounting accreditation process. The AAA was in the middle of this controversy and was developing its own accreditation standards for submission to some accrediting body. Initially, the AAPG simply watched the accreditation debate and summarized the activities for the benefit of members. In late 1978, AAPG President Fred Skousen wrote the AACSB requesting that the group be permitted to have an observer at committee meetings relating to accreditation of accounting programs. That request was denied ["Summary...," 1979, p. 2]. When William Markell became president in 1979, he eliminated the accreditation committee, but this was because the board of governors incorporated the accreditation issue under its domain. The Fall 1979 issue of the newsletter was devoted to accreditation issues. When the AACSB issued its proposed standards in the summer of 1979, the board went on record as unanimously opposing any standards for baccalaureate programs in accounting. It was stated, "to accredit four year baccalaureate programs would be a step backward." The argument was that the long-term goal should be to raise educational standards for the accounting profession and to accredit baccalaureate programs would only delay the implementation of a five-year requirement ["Board...," 1979, p. 7]. The position of the AAPG was rejected by the AACSB.

Accreditation has remained a concern of the AAPG throughout the group's history. The Accreditation Committee was reinstituted after a one-year hiatus and was quite active during the early 1980s. From 1982 through 1984 the issues of concern included the lack of representation of accounting administrators in the AACSB accreditation process and the possibility for accrediting accounting programs at schools where the business school could not meet the accreditation standards. In 1985 and 1986 there were joint discussions with officers of the Federation of Schools of Accountancy (FSA) regarding either getting more input into the AACSB or forming an organization to replace the AACSB. The basic conclusion was that letters should be written to the AACSB expressing concern about the status of accounting accreditation at the AACSB [Minutes of Meeting, February 11, 1985, p. 5]. To further accentuate the importance of accreditation in the activities of AAPG, the title of the chairman of the accreditation committee was changed to vice president of accreditation in 1988. James J. Benjamin was the first vice president of accreditation (1988-89), followed by Gary E. White (1989-90) and Robert G. May (1990-91).

Research and Data Base Committees

The Research and Publications Committee started off enthusiastically. Its first recommendation, made at the April 1977 board of governors meeting, was that the group should publish a journal, tentatively entitled "Journal of Accounting Education." The Board asked the committee to come back with a plan to underwrite the costs of such a journal. For 1977-78, the research and publication areas were divided between two committees. The new Publications Committee was divided as to whether a new journal was needed, or more pages devoted to educational research in *The Accounting Review*. The aforementioned financial plan was not submitted and the Board sent the issue back to the committee for further review. The two areas of publications and research were again reunited for 1978-79, and the emphasis of that committee was the implementation of a monograph series and the continuation of the annual faculty market survey conducted by Bradley University professors Arthur Mehl and Lucille Lammers. The subject of a journal was not discussed in the committee's report. However, it is interesting to note that at its November 1978 meeting, the AAA Executive Committee discussed the possibilities of the Association (not AAPG) establishing a journal entitled "Journal of Accounting Education," but no action was taken at that time. Subsequently, the School of Accountancy at James Madison University began publishing its own *Journal of Accounting Education*. That journal, combined with the later *Issues In Accounting Education* published by the AAA filled the niche for publication outlets for educational research.

The promoting of research of interest to accounting administrators has probably been the AAPG's weakest area of activity if the program for the annual AAA meeting is any indication. The acceptance rate for papers at the annual meeting has been about 22% over the past couple of years, but almost no papers have been submitted for the AAPG-sponsored session. As a result, the session allotted to the AAPG has ended up being a panel discussion due to a lack of submissions.

At the first board of governors meeting in October, 1977, the board voted to distribute a quarterly newsletter to members. That idea was subsequently cut back to a semiannual newsletter. The first issue was dated Winter (January), 1977, under the editorship of Vincent Brenner. Subsequent editors are listed in Exhibit 5-1. The editor also served as the chairman of the Research and Publications Committee.

One of the first major votes of the membership was taken at the 1977 annual meeting in Portland. The AAA Executive Committee had passed a provision which would ban recruiting activities during luncheons and plenary sessions at future annual meetings of the Association. This included closing the placement room. After heated debate, the 125 members in attendance voted overwhelmingly against the proposal; AAPG president Doyle Williams was asked to communicate the results to the Executive Committee. The Executive Committee did not listen to the AAPG, and the ban was put into effect.

Previously, it was mentioned that the AAPG supported the annual faculty study conducted by Mehl and Lammers, but that study was considered insufficient for the needs of many administrators. Thus, for 1978-79, a Data Base Committee, under the leadership of Charles G. Carpenter (Miami University), was established to compile salary and other data that could prove useful to accountancy administrators. The following year's committee, still chaired by Carpenter, had as its charge the development and implementation of a system for collecting, distributing, and maintaining a national data base for accounting programs. Doyle Williams (University of Southern California) was named as the data base project director. Since 1979, a questionnaire has been sent out each autumn requesting data on faculty salaries.

Financial Considerations

The AAPG has never been restrained in its activities by a lack of funding. Most years have ended with an excess of revenues over expenditures (causing at least one commentator to observe that administrators are so accustomed to saying "no" to expenditure requests that they continue to do so even when funds are available). The 1977 fiscal year ended with an excess of $1,141 on revenues of $10,564.

At the February 1978 board meeting, President Williams announced the receipt of the Group's first grant; Touche Ross & Co. had donated $5,000 to cover the start-up costs of the AAPG. As a result, the AAPG had a cash balance of $9,635 at the end of August 1978. Despite a proposed $2,000 budget deficit for 1979, that fiscal year ended with an excess of receipts over expenditures of $6,471 (on revenues of $11,325), bringing the ending cash balance to $16,106. The lack of expenditures was attributable to several committees not spending the funds which had been budgeted for their use. By the end of 1980, the Group had $17,183 in cash, and that increased to $21,678 by August 31, 1981. The 1981-82 budget called for a deficit of $6,700. However, the deficit was again avoided, and the August 1982 balance was $23,506. At the February 1982 board of governors meeting, Gary White (Texas Tech) questioned whether perhaps the dues were too high since the Group was apparently unable to expend all of the money it was receiving. The remainder of the board, however, assured White that dues were not too high.

The 1982-83 officers worked more diligently at spending money, and for the first time ever a deficit was achieved. At August 31, 1983, the cash balance had been reduced to $18,374. There had been a new expenditure during the year which helped achieve the long-awaited deficit—namely the costs of the new regional vice presidents who were responsible for holding meetings in their respective regions.

Another deficit was budgeted for 1983-84, this one for $10,500. That level was nearly achieved as the August 1984 balance was down to $11,160. A deficit of $7,253 was budgeted for 1984-85, but every line item came in under budget and a slight surplus was achieved for the year. The 1985 ending balance was $12,096.

Total dues revenue for the year had been $12,525 as there had been little change in membership over the years. In fact, membership has remained relatively stable throughout the history of the organization, and to this date, there has never been a dues increase. There was a discussion of a dues increase in early 1985, but the officers were unsure how such a matter could be implemented because the by-laws stipulated that an increase in dues required the vote of the membership.

The AAPG continued to be a profit center during the late 1980s. By August 1988, the cash balance was up to $21,985. The 1989 fiscal year closed with $23,760. This latter surplus of $1,775 was based on a budget which called for an $11,000 deficit. A deficit of $7,750 was called for in the 1989-90 budget, but due to a lack of spending by the regional vice presidents, the actual deficit was only $3,227, resulting in a June 30, 1990 ending balance of $20,532. A new fiscal year end was approved at the August 9, 1990, board of governors meeting. That new year end was retroactive to the preceding June 30. The primary reason for the change was to provide a smoother transition of financial reporting responsibilities, especially reporting to the annual AAPG business meeting in August.

AAPG Summary

Fifteen years of AAPG activities can be summarized in a few sentences. The Group spent a great amount of time and resources in its early years on the subject of accreditation. Although little was accomplished by these endeavors, the AAPG has been a vocal group looking after the interests of its membership. Probably the most successful activity of the AAPG has been the annual seminars held each winter. A high proportion, often 2/3, of the membership regularly attends these seminars. The data base project has been another area of success. A major service is supplied to members through these annual survey results.

In many ways the AAPG has not been as productive as the AAA sections have been; there is no journal, monograph series, or annual awards of any kind. Obviously, there is room to expand activities. On the other hand, the AAPG is hampered by limited membership and that membership is constantly fluctuating due to turnover among administrators. Unlike the sections, the AAPG does not have members who have a special interest in the subject matter of the section. Thus, given the small and often temporary membership, and the limited time available to that membership, the AAPG cannot be faulted for its level of activities over the years. In many respects the AAPG should be lauded for what it is accomplishing with so little in the way of resources. Another consideration is that members do not want too much from the organization other than the opportunity to mix with others of their own kind and share experiences. Accountancy chairmen, directors, and deans need others who can sympathize with the problems faced by an administrator. Surely the success of the seminars and the high attendance at the annual meeting is indicative of this need for peer companionship.

AUDITING SECTION

The Auditing Section was one of the first two sections approved by the Executive Committee at the March 1976 meeting. Its purpose is to give greater attention to the area of auditing. Objectives are divided into Education, Research, and Profession categories. The Auditing Section was a fast growing section. By July 1980, Auditing had 1,072 members, which was more than 240 greater than the American Taxation Association membership, and 500 greater than any of the remaining six sections. Fred Neumann (University of Illinois) was the founding father of the Auditing Section and served as its first chairman. Alvin A. Arens (Michigan State University) was the first vice chairman, while Jack Robertson (University of Texas) served as the first secretary. Exhibit 5-2 lists the officers for later years. In addition to the aforementioned officers, the position of vice chairperson from practice was added in 1979 with James Loebbecke as the first incumbent. This latter position was changed to one with a two-year tenure in 1988 and the election was staggered with that of the secretary, which is also a position with a two-year tenure. For most chairmen, the section position has been a stepping stone to bigger things as several have gone on to serve on the AAA Executive Committee.

Unlike most of the other AAA sections, the Auditing Section's nominating committee normally nominates two individuals for each vacant position. This policy is the result of a referendum passed at the 1978 business meeting in Denver. The

Exhibit 5-2
Auditing Section Officers

Year	Chairman	Vice Chairman	Practice Vice Chairman	Secretary	News Editor
1977	Frederick Neumann	Alvin A. Arens		Jack C. Robertson	
1978	Alvin A. Arens	Jack C. Robertson		Donald A. Leslie	Corine T. Norgaard
1979	Jack Robertson	Neil Churchill	James Loebbecke	Jay M. Smith, Jr.	Corine T. Norgaard
1980	Neil Churchill	Bart H. Ward	James Loebbecke	Fred Davis	Jack Krogstad
1981	Bart Ward	William Felix, Jr.	James Kirtland	Donald R. Nichols	Jack Krogstad
1982	William Felix, Jr.	James Loebbecke	James Kirtland	Jack Krogstad	Gerald Smith
1983	James Loebbecke	Joseph Schultz	John Willingham	Jack Krogstad	Gerald Smith
1984	Joseph Schultz	Jack Krogstad	John Willingham	Wanda Wallace	Gerald Smith
1985	Jack Krogstad	Gary Holstrum	Lynford E. Graham	Gerald Smith	Arnold Wright
1986	Gary L. Holstrum	D. Dewey Ward	Lynford E. Graham	Gerald Smith	Arnold Wright
1987	D. Dewey Ward	Gerald Smith	Donald A. Leslie	Carl S. Warren	Dennis L. Kimmell
1988	Gerald Smith	Robert May	Robert S. Roussey	Jean C. Wyer	Dennis L. Kimmell
1989	Robert May	Andrew D. Bailey	Robert S. Roussey	Jean C. Wyer	Dennis L. Kimmell
1990	Andrew D. Bailey	Carl S. Warren	Robert S. Roussey	Zoe-Vonna Palmrose	Wanda Wallace
1991	Carl S. Warren	Theodore Mock	Brent C. Inman	Zoe-Vonna Palmrose	Wanda Wallace

procedure was readdressed in 1985 because some officers felt that the practice of having contested elections caused some people to decline a nomination. By a vote of 55 to 36 the Section voted to continue the previous policy.

Publications

The section has published a triennial newsletter, entitled *The Auditor's Report*, from its earliest days. In addition to news items, the newsletter often contains short articles on auditing-related subjects. Editors are listed in Exhibit 5-2.

At its annual meeting in Boston in 1980, the Auditing Section membership approved the recommendation of its Auditing Journal Task Force that two issues of a section journal be published in 1981. William W. Cooper (Harvard University) was appointed as the first editor of *Auditing: A Journal of Practice & Theory*. Carl S. Warren (University of Georgia) followed from 1981-84. From 1984 through 1987, Andrew Bailey and Kurt Pany (Arizona State University) served as coeditors. They were followed by Jack Krogstad (Creighton University) and most recently by Joseph J. Schultz (Arizona State University).

The journal received a significant financial boost when Don Leslie, Albert Teitlebaum, and Rod Anderson were awarded the 1980 Wildman Medal for their book *Dollar-Unit Sampling: A Practical Guide for Auditors*. Section members Don Leslie and Rod Anderson, partners with the CPA firm of Clarkson Gordon, requested that the AAA give their share ($1,667) of the $2,500 award to the Auditing Section for support of the new journal. Section chairman Bart Ward indicated that the funds would be used to underwrite a major portion of the production costs associated with the first issue of the journal ["Auditing Section...," 1981, p. 12]. The journal is considered to be a semiannual publication, but occasional special issues are published containing the proceedings of auditing symposia cosponsored by the section.

Auditing Awards

The Auditing Section inaugurated a distinguished service award in 1983. Recipients of this award have been:

1983	Kenneth Stringer
1984	Robert Mautz
1985	Robert K. Elliott (Peat, Marwick, Mitchell & Co.)
1988	W. W. Cooper (University of Texas)
1989	James K. Loebbecke (University of Utah)
1990	Donald A. Leslie (Clarkson Gordon)

In addition to the outstanding service award, the section also gives an outstanding educator award, which has gone to the following individuals:

1988   Alvin A. Arens (Michigan State University)
1989   William L. Felix (University of Arizona)
1990   Howard F. Stettler (University of Kansas)

The section began giving an outstanding dissertation award in 1988. The first winner was Mary T. Washington (University of Southern California) for her dissertation entitled "Audit Evidence Evaluation as a Cascaded Inference Task."

Auditing Section Finances

The Auditing Section has typically been a profit center. The section ended the 1981 fiscal year with $11,929. A deficit in 1982 left $7,608. In 1983, the cash balance increased by $4,262 to $11,870. That amount increased to $12,734 in 1984. This represented a reserve of nearly a year's dues since both 1983 and 1984 showed dues collections at under $15,000. Another deficit was incurred in 1985 leaving a $6,016 balance, but in 1986 the cash was back up to $12,787. In 1987 a grant was received from Peat, Marwick, Main for the Research Agenda Project; the result was a 1987 surplus of $12,787, and a cash balance of $32,814. The section closed out 1988 with a $41,179 cash balance, but that declined to $30,638 in 1989 (1989 covered only ten months due to a change in fiscal year). The major reason for the 1989 deficit was the expenditures for the Research Agenda Project.

Auditing Section Summary

As one of the oldest AAA sections, and the largest section, the Auditing Section has a rich history. It was the second section to publish a journal, and that journal has engendered an excellent reputation. The section has long had an outstanding newsletter; it has sponsored symposia, gotten practitioners involved with academicians, sponsored luncheons and other programs at national meetings, and has been active at regional meetings. In other words, the Auditing Section has done it all. The Auditing Section can serve as a model for other sections.

GOVERNMENT AND NONPROFIT ACCOUNTING SECTION

What now goes by the name Government and Nonprofit (G&NP) Section was originally established in 1976 as the Public Sector Section. Along with the Auditing Section, the Public Sector was one of the first two AAA sections to be approved by the Executive Committee at its March 1976 meeting. From the beginning, the members of the section have represented a broad array of nonprofit organizations including federal, state, and local governments, hospitals, colleges and universities, public schools, and voluntary health and welfare organizations. Objectives of the section are (1) to foster basic and applied research in financial and managerial accounting, as well as auditing, for governmental and nonprofit organizations; and (2) to improve the quality of accounting, financial reporting, and auditing in these organizations through advances in teaching, research, and service.

The organizational meeting of the Public Sector Section was held in Atlanta in August 1976. Leon E. Hay (Indiana University) was elected as the first chairman of the section, Robert Freeman (University of Alabama) was vice chairman, William A. Broadus (GAO) was secretary, and William W. Holder (Texas Tech University) was editor of the newsletter. National officers for subsequent years are listed in Exhibit 5-3.

One reason that the section was established was because individuals teaching and conducting research in the area of governmental and nonprofit accounting had a feeling that they were being ignored by the mainstream accounting educators. The charges to the initial education committee are indicative of the concerns of the founders of the section. Those charges were to:

1. Attempt to ensure inclusion of public sector topics in program for 1977 AAA Doctoral Consortium.

2. Maintain liaison with professional organizations concerned with public sector accounting and encourage coverage of public sector topics in programs of national and regional meetings of such organizations.

3. Facilitate distribution of course syllabi, teaching aids, and study materials by direct exchange and/or by publication.

Exhibit 5-3
Governmental and Nonprofit Accounting Section Officers

Year	Chairman	Chair Elect	Secretary	News Editor
1977	Leon E. Hay	Robert J. Freeman	William A. Broadus	William W. Holder
1978	Robert J. Freeman	Emerson O. Henke	William A. Broadus	William W. Holder
1979	Emerson O. Henke	William A. Broadus	William W. Holder	Robert W. Ingram
1980	William A. Broadus	William W. Holder	Michael H. Granof	Robert W. Ingram
1981	William W. Holder	Michael H. Granof	Abraham J. Simon	Mortimer A. Dittenhofer
1982	Michael H. Granof	Abraham J. Simon	Robert W. Ingram	Mortimer A. Dittenhofer
1983	Abraham J. Simon	Robert W. Ingram	Mortimer A. Dittenhofer	Leonard E. Berry
1984	Robert W. Ingram	Mortimer A. Dittenhofer	Leonard E. Berry	Gary Giroux
1985	Mortimer A. Dittenhofer	Leonard E. Berry	John Engstrom	Gary Giroux
1986	Leonard E. Berry	John H. Engstrom	Wanda A. Wallace	R. Penny Marquette
1987	John H. Engstrom	Wanda A. Wallace	James L. Chan	R. Penny Marquette
1988	Wanda A. Wallace	James L. Chan	R. Penny Marquette	Jesse Hughes
1989	James L. Chan	R. Penny Marquette	James M. Patton	Jesse Hughes
1990	R. Penny Marquette	James M. Patton	Jesse Hughes	Penny Wardlow
1991	James M. Patton	Jesse Hughes	Earl Wilson	Penny Wardlow

The committee was successful in its efforts; there were public sector sessions at all of the AAA regional meetings, and a session on social audits at the Doctoral Consortium. The publication of education committee findings was facilitated by a column which began in the section newsletter in May 1977 entitled "Teaching Review." Liaison with other groups was evidenced by the cosponsorship with the Association of Government Accountants (AGA) of three workshops on education of government accountants, which were held in Washington, DC in June 1977. The same charges were given the committee the following year.

The charges to the initial research committee were also three in number:

1. Develop an annotated list of public sector basic and applied research topics and disseminate that list.

2. Propose to AAA President-Elect that there be national research committees for 1977-78 that will consider public sector topics.

3. Solicit and evaluate research reports on public sector topics submitted by section members and others and arrange for publication of meritorious reports in working paper form or as AAA monograph.

The annotated list of research topics (nine pages in length) was prepared and distributed to members attending the 1977 annual meeting in Portland.

The first consideration of a name change for the organization arose at the 1979 annual meeting in Hawaii. Comments were made that the name Public Sector Section was vague, confusing, and frequently interpreted in a narrow fashion. No motion was forthcoming, but Chairman Emerson Henke requested that the problem be noted in the newsletter in an effort to elicit other names that might provide a clearer indication of the nature and scope of the section. No names were received during the ensuing year. In early 1984 a membership survey was conducted. Although the response rate was only five percent, those who did respond favored a name change.

One result of the aforementioned survey was an indication that not enough emphasis was being given to nonprofit accounting. Indeed, this was a valid criticism since nongovernmental accounting had scarcely ever been mentioned in the newsletters published during the first eight years of the section's history. Given this lack of emphasis on nonprofit organizations, the eventual name change which came in August 1984 was somewhat surprising since it seems that limiting the organization strictly to governmental accounting would have been the logical alternative (after all, it had been limited to governmental accounting for eight years). One of the main reasons for the change was that the section was being confused with the Public Interest Section which had been formally established in 1982.

Publications

The G&NP Section began publishing a quarterly newsletter in December 1976. Admittedly, some of the early newsletters were on a single sheet of 8-1/2-by-11 paper, but regardless of thickness, there has always been a timely newsletter. The newsletter typically reports on section operations, including committee activities, plus includes occasional book reviews and the availability of working papers. The title of the newsletter was changed to *Public Sector Section News* beginning with the spring 1982 issue. With the change in the name of the section, the title of the publication was changed to *Government & Nonprofit News* effective with the fall 1985 issue. Surprisingly, the printing quality of the newsletter has deteriorated over the years. Whereas other sections started out with crude newsletters and eventually switched to more impressive versions, the G&NP Section began with a beautifully printed publication on slick paper, which even allowed for excellent reproduction of photos. Beginning with the spring 1982 issue, the slick paper was eliminated, but still the printing was of high quality. In 1984, both the quality of the paper and the printing declined as the newsletter appeared to be prepared on a typewriter or computer printer and then copied. The quality improved again in 1989 when the AAA headquarters adopted a new desktop publishing system.

The working paper series was begun in late 1978 under the editorship of research committee chairman Abraham J. Simon (Queens College, CUNY). Because of limited section funds, authors had to provide 25 to 50 copies of their papers if they wanted them distributed. Beginning in 1981, abstracts were published in a semiannual supplement to the newsletter. That supplement was entitled *Abstracts of Public Sector Research.*

At the 1981 annual meeting in Chicago, there was consideration of a proposal for the section to cosponsor a new journal, *Public Budgeting and Finance.* Research Committee chairman Robert Ingram proposed that the section join with the American Society for Public Administration, the American Association for Budget and Program Analysis, and George Mason University to sponsor the publication. Cosponsorship would require the section to contribute $10 per member per year. Although it was acknowledged that this would require a dues increase of $10 per year, the section would be in a position to influence the content of the journal. The ensuing discussion was quite controversial. Some members wanted to add the word "accounting" to the title of the journal. Others questioned whether the section would really be able to influence content. The final action taken was to appoint a committee to poll the membership regarding whether they were willing to pay the extra $10 a year, and, if that were acceptable, to work out an arrangement with the other cosponsors to ensure that the journal would include public sector accounting as one of the three major topic areas to be covered in its contents ["Section Annual...," 1981, pp. 1-2]. A year later, Michael Granof reported that the section had decided not to become a cosponsor, partially because the publishers had little interest in adding another sponsor ["Section Conducts....," 1982, pp 4-5].

The section did publish a monograph in 1982, *Research in Government Accounting, Reporting, Budgeting and Auditing: Evaluation and Issues for Research*, which was authored by the 1981-82 research committee. The firms of Touche Ross and Deloitte Haskins & Sells underwrote the publication costs. A year later the 600 copies that had been printed initially were gone and the Chicago office of Peat Marwick Mitchell & Co. funded a second printing. Another monograph was published in 1986 entitled *An Annotated Bibliography of Articles in Governmental Accounting, Auditing and Municipal Finance: 1970-1985*. This publication, edited by Penny Marquette (University of Akron), abstracted over 1,000 articles from 31 different journals. In 1989, two publications were issued: *Measuring the Performance of Nonprofit Organizations*, a 341-page report by the Committee on Nonprofit Entities' Performance Measures, chaired by Teresa Gordon (University of Idaho), and *Cases in Government and Nonprofit Accounting*, developed by the Education Committee chaired by Walter Robbins (University of Alabama). In addition to these research-type publications, the section has also published membership directories in various years. There was also a 1986 *Regional Coordinators' Handbook*, which was compiled by Penny Marquette.

G&NP Section Financial Affairs

During its first year (through June 30, 1977), the G&NP Section collected $2,043 in membership dues (indicating about 204 members). Expenditures were kept to a minimum and the ending cash balance stood at $1,286. The section never had an overabundance of money as the newsletter used up most funds. For instance, in 1981-82, the expenditures for the newsletter exceeded the receipts from dues. As a result, the section ended the 1982 fiscal year with only $338. That figure had increased to $1,200 by the end of 1983, and because of a change in the method of printing the newsletter (lower quality), there was a substantial surplus for 1984 resulting in an ending cash balance of close to $3,700 ["Section Annual....," 1984, p. 2]. The cash had increased to $9,100 by the end of 1985, a year in which the $1,700 expenditure for the newsletter was one-third what it had been three years earlier. The section continued to be a profit center in 1986 and 1987 as those fiscal years ended with cash balances of $10,682 and $14,594, respectively. The fund balance had increased to $16,625 by the end of 1988, and $21,581 in 1989. The $4,956 surplus in 1989 was in comparison to a budgeted deficit of $5,300.

Awards

The G&NP Section inaugurated an annual dissertation award in 1984. Dissertation award winners were:

1984— Susan Herhold Baskin, "The Information Content of the Audit Report as Perceived by Municipal Analysts."

1985— Marc A. Rubin (University of Texas), "An Examination of the Political and Economic Determinants of Audit Fees: Theory and Evidence."

1986— Stephen D. Willits (Texas Tech University), "Public Employee Retirement Systems Reports: A Study of User Information Processing Ability."

1988— Paul A. Copley (University of Alabama), "An Empirical Investigation of the Determinants of Local Government Audit Fees."

1989—(tie) Sharon Green (University of Pittsburgh), "A Behavioral Investigation of the Effects of Alternative Governmental Reporting Formats on Analysts' Predictions of Bond Ratings."

1989—(tie) Carol Lawrence (Indiana University), "The Effect of Accounting System Type and Ownership Structure on Hospital Costs."

In 1986, the section membership approved an award for lifetime achievement in the area of governmental or nonprofit accounting. The first such award was presented to Robert Anthony (Harvard University) in 1988 for his enduring contributions to the profession over the past 40 years. Mortimer Dittenhofer (Florida International University) was granted the award in 1990.

G&NP Section Summary

The G&NP Section currently has about 700 members. In 1976, its dues were a rather high $10 per year, but those dues were never increased through 1990, and are now among the lowest dues of any AAA section. The section holds the distinction of having issued more newsletters than any other AAA section since it is the only section to have been in business for 15 years and have published a quarterly newsletter throughout that time. Besides the newsletter, the major activities of the section during the first eight years was involved with getting public sector sessions on the programs at regional and national AAA meetings. At this endeavor it was quite effective. Still, the section has not been noted for its aggressive behavior. As newsletter editor Mortimer Dittenhofer noted in 1982:

> We seem to be a passive type of organization. The character of the Section belies the actual character of many of our members who are aggressive and innovative [Dittenhofer, 1982, p. 4].

Following the name change in 1984, the activities of the section increased. At about the same time the resources of the section were transferred from its newsletter to other activities such as awards and increased committee endeavors.

Earlier it was mentioned that the tax professors who founded the ATA felt that taxation was the orphan of the accounting field. The members of the G&NP Section would probably argue that statement with the feeling that if anyone is misunderstood, it is those who teach governmental and nonprofit accounting. Indeed, 1981-

82 section chairman Michael Granof (University of Texas) may have best summarized why the section has not grown as large as those devoted to other fields of interest:

> To state that programs in nonprofit accounting lack the prestige of those in related areas of accounting may be to belabor the obvious. In recent years other specialties in accounting—auditing, tax, systems and consulting—have become more glamorous and appealing to students seeking high-paying and challenging fields in the business domain. By contrast, public sector accounting seems to be associated still with government bureaucracy and all its dreary trappings [Granof, 1982, p. 2].

If even the section chairman feels this way, is it any wonder why the activities of the section have not been more exciting.

INFORMATION SYSTEMS/MANAGEMENT ADVISORY SERVICES SECTION

The Management Advisory Services (MAS) Section was approved by the Executive Committee at the August 1976 meeting in Atlanta. Lynn J. McKell (Brigham Young University) was the first chairman, and was assisted by vice chairman Charles Litecky (University of Missouri). Officers for later years are listed in Exhibit 5-4.

At the March 1980 AAA Executive Committee meeting, a request from John T. Overbey was discussed regarding the establishment of an Accounting Information Systems/Management Information Systems Section. Several members of the Executive Committee questioned the subject content of the section and its possible overlap with the MAS Section. Overbey's request was tabled with instructions for him to provide more information regarding the purposes and subject content of the

Exhibit 5-4
IS/MAS Officers

Year	Chairman	Vice Chairman	Secretary	News Editor
1977	Lynn J. McKell	Charles Litecky	Gary Grudnitski	Thomas E. Gibbs
1978	Charles Litecky	John O. Mason, Jr.	Gary Grudnitski	Thomas E. Gibbs
1979	John O. Mason, Jr.	Edward L. Summers	Thomas E. Gibbs	Myles Stern
1980	Edward L. Summers	Thomas E. Gibbs	Rick Elam	Myles Stern
1981	Thomas E. Gibbs	A. Douglas Hillman	Donald E. Bostrom	Myles Stern/C. Litecky
1982	A. Douglas Hillman	James C. Kinard	J. Owen Cherrington	Charles Litecky
1983	James C. Kinard	J. Owen Cherrington	Joseph W. Wilkinson	Charles Litecky
1984	J. Owen Cherrington	Gary Grudnitski	Joseph W. Wilkinson	Charles Litecky
1985	Gary Grudnitski	William E. McCarthy	Marshall B. Romney	Sue H. McKinley
1986	William E. McCarthy	Marshall B. Romney	Clinton E. White	Sue H. McKinley
1987	Marshall B. Romney	Joseph L. Sardinas	A. Faye Borthick	Severin V. Grabski
1988	Joseph L. Sardinas	A. Faye Borthick	Graham Gal	Severin V. Grabski
1989	A. Faye Borthick	Graham Gal	Severin V. Grabski	Kevin Stocks
1990	Graham Gal	Severin V. Grabski	Casper E. Wiggins	Kevin Stocks
1991	Severin V. Grabski	Casper E. Wiggins	Anita S. Hollander	Ronald L. Clark

proposed section [Minutes, March, 1980, p. 10]. Three years later, the name of the section was officially changed to Information Systems/Management Advisory Services (IS/MAS) Section following a positive vote by an overwhelming majority of the section membership [Minutes, March 1983, p. 15].

The section's publications have included a newsletter, the editors of which are listed in Exhibit 5-4. In addition, a semiannual journal was established in the fall of 1986 under the editorship of Joseph Wilkinson (Arizona State University). William McCarthy (Michigan State University) took over the editorship of *The Journal of Information Systems* beginning with the fall 1989 issue.

The IS/MAS Section has grown to over 900 members. Although its committee activities have not been as great as that of many of the other sections, it has been active at the regional level. In addition to the AAA regions, the section maintains a program liaison with the MAS Division of the AICPA. In summary, the IS/MAS Section provides an opportunity for cohesiveness for individuals whose interests are outside the mainstream of most AAA members, and at that it has been quite successful. Toward that end, the section provides forums for presentation and discussion of systems and MAS subjects, and publishes a journal and newsletter for section members. It is the smallest of the five sections which publish journals.

TWO-YEAR COLLEGE SECTION

One of the five original sections was the Community/Junior College Section which held its first meeting in Atlanta in August 1976. Joe Rhile (Lake-Sumter Community College, Leesburgh, FL) was the first chairman. Jack Topiol (Community College of Philadelphia) was vice chairman and Albert T. Pasek (Lake Michigan Community College, Benton Harbor, MI) was secretary. Later year's officers are listed in Exhibit 5-5. Dues in 1976 were set at $5 per year (which along with the MAS Section was the lowest of the five original sections). The objectives of the section are to gather and promulgate information pertaining to accounting education in two-year colleges and to stimulate advances in the teaching of accounting courses.

A newsletter, *The Quarterly*, is issued three times per year (which might cause some to question why it is called *The Quarterly*. Beyond the newsletter, the section is not particularly active. It does promote the submission of two-year-college papers to the AAA annual meeting, but it has not been successful in this regard. The section was active in the early 1980s in sponsoring microcomputer seminars at annual meetings, and in that regard led the field in

Joe Rhile

Exhibit 5-5
Two-Year College Section Officers

Year	Chairman	Vice Chairman	Secretary & News Editor
1977	Joe Rhile	Jack Topiol	Albert T. Pasek
1978	Jack Topiol	Albert T. Pasek	Lorraine Hicks
1979	Albert T. Pasek	Lorraine Hicks	Lee C. Wilson
1980	Lorraine Hicks	Lee C. Wilson	Melissa Martinson
1981	Lee C. Wilson	Melissa L. Martinson	Herman Andress
1982	Melissa L. Martinson	Herman Andress	Susan Harrison
1983	Herman Andress	Susan H. Harrison	Thelma Mitchell
1984	Susan H. Harrison	Thelma Mitchell	Christopher J. Trunkfield
1985	Thelma Mitchell	Christopher J. Trunkfield	Robert J. McCloy
1986	Christopher J. Trunkfield	Jerry A. Van Os	Joseph A. Kreutle
1987	Jerry A. Van Os	Joseph A. Kreutle	Jacqueline Sanders
1988	Joseph A. Kreutle	Jacqueline Sanders	Beverly Boyce Terry
1989	Jacqueline Sanders	Beverly Boyce Terry	Helen Gerrard
1990	Beverly Boyce Terry	Helen Gerrard	Billie Cunningham
1991	Helen Gerrard	Billie Cunningham	Pat Reihing

such activities. The biggest contribution made by the section has been the publication of *The Two-Year College Faculty Directory*, which is compiled annually since 1986 by Joe Rhile and published by South-Western Publishing Company. Prior to Rhile's book, there was no authoritative list of two-year college teachers of accounting.

Today, the Two-Year College Section has about 500 members and serves a niche in the accounting education marketplace that is often overlooked by other sections and organizations.

ACCOUNTING, BEHAVIOR AND ORGANIZATIONS SECTION

The Accounting, Behavior and Organizations (ABO) Section, which now has over 1,200 members, was granted provisional status at the March 1981 meeting in San Diego of the Executive Committee (by a vote of 7-to-1). By March 1982, ABO had 224 members and the Executive Committee granted the organization full sectional status. The first section chairman was Gerald H. B. Ross (University of Michigan). Eric Flamholtz (UCLA) was the first vice chairman, Van Ballew (San Diego State) was the first secretary-treasurer, and Ellen Cook (University of San Diego) was the newsletter editor. This group served from the time the organization received provisional status in 1981 through 1983. Officers for later years are listed in Exhibit 5-6. Stephen Landekich of the National Association of Accountants was the first to hold the practitioner position in 1984-85. The position was left unfilled from 1988-1990 due to difficulty in finding individuals to fill the position. In 1990, a bylaws amendment changed the term of the vice president-practice from one to two years. The reason for the change was to allow the practitioner more time to be integrated into the activities of the section.

The section was formed to address two related fields. The first is the interface between the behavioral sciences and accounting. The second is the link between organizational theory and accounting. The objectives of the section are (1) to stimulate interchange and research on the relationship between behavioral science and accounting, and (2) promote the integration of new constructs of organizational effectiveness, as elaborated in organizational and general system theories, with developments in accounting theory [Ross, 1982, p. 1].

Publications

The ABO Section's history of publishing a newsletter began two years before volume 1, number 1 was issued in 1984. That anomaly requires some explanation. Even though the ABO newsletter as it is now known was established in the spring of 1984, there had been two years of newsletters with the same title prior to that time. Apparently the reason for designating the 1984 newsletter as volume 1 was because the new version was professionally produced and more attractive than the earlier version. Ellen Cook edited a semiannual *ABO Newsletter* beginning with a May 1982 issue. These early newsletters were typed and then copied on thin paper on a copying machine. The University of San Diego financed the newsletters for the first two years. By February 1984, the section had over 400 members, and there were by then funds available for a better quality, printed, newsletter [Holtfreter, 1984, p. 2]. The spring 1984 newsletter had a format similar to that of today's newsletter. Through 1985, the two semiannual issues were for some reason dated "spring" and "summer," although judging from the content, the spring issue was apparently written during the preceding autumn. A spring and fall schedule was adopted in 1986, but this was changed to summer-winter in 1987.

The first issue in the new format was ten pages in length and included such newsworthy items as behavioral papers presented at the AAA annual and regional meetings and at a meeting of the American Institute for Decision Sciences. There

Exhibit 5-6
ABO Section Officers

Year	Chairman	Vice Chairman	Practice Vice Chairman	Secretary	News Editor
1982	Gerald H. B. Ross	Eric Flamholtz		Van Ballew	Ellen Cook
1983	Gerald H. B. Ross	Eric Flamholtz		Van Ballew	Ellen Cook
1984	Robert E. Holtfreter	Ellen Cook		Ray Stephens	Joe San Miquel
1985	Ellen Cook	Ray G. Stephens	Stephen Landekich	Frank Collins	Peter Chalos
1986	Ray G. Stephens	Frank Collins	James C. Caldwell	Martin Bariff	Peter Chalos
1987	Frank Collins	Martin Bariff	Michael E. Egan	Kenneth Merchant	Peter Chalos
1988	Martin Bariff	Kenneth Merchant	Richard Sabo	Norman MacIntosh	Paul H. Munter
1989	Kenneth Merchant	Norman MacIntosh		Kenneth Ferris	Paul H. Munter
1990	Kenneth Ferris	Howard Rockness	Mike Gleason	Mark Haskins	Paul H. Munter
1991	Howard Rockness	Mark Haskins	Mike Gleason	Paul H. Munter	Don W. Finn

was also an annotated bibliography of articles appearing in non-accounting journals which might be of interest to members. Behavioral articles in upcoming issues of *Decision Sciences* and *Accounting, Organizations and Society* were listed as were working papers submitted by section members. The newsletter has never been particularly newsy with respect to section operations. There has never been a financial statement published, and until the winter 1991 issue there had never been a printing of the minutes of officer or annual business meetings.

A working paper series has been an aspect of the ABO Section from its founding. The first working paper series editor was Pekin Ogan (Indiana University) who provided an excellent service in not only soliciting working papers, but also in reporting and disseminating them to members. Since then, every issue of the newsletter has listed working papers that could be obtained by writing to the authors. In fact, a high percentage of the newsletter has been devoted to working papers in recent years. In the winter 1991 issue, descriptions of working papers accounted for nine of the 20 non-advertising pages.

Other publications sponsored by the ABO Section have included a compilation of behavioral accounting course syllabi, prepared by Robert Holtfreter (1986), and an annotated bibliography of behavioral accounting publications, which was compiled by Frank Collins and Don W. Finn, both of Texas Tech University, in 1986. Both the syllabi and the bibliography were published after several delays and postponements. The bibliography was subsequently updated in 1988 and made available free to members in a DBase III format. The behavioral accounting syllabi publication was updated in 1990 under the editorship of Penelope Sue Greenberg and Ralph Greenberg, both of Temple University. The 1990 version (181 pages) contained syllabi from 24 accounting courses where instructors had incorporated behavioral and organizational concepts into their classes.

A Section Journal

The idea of a section journal was discussed almost from the founding of the section, but it was not until the summer of 1984 that a task force, chaired by Joe San Miguel, was appointed to study the prospects for a new journal. The initial report suggested deferring the idea until a later time. One reason for the deferral was because *Accounting, Organizations and Society (AOS)* was increasing its page count, which would allow more space for behavioral articles. There was even consideration of adopting *AOS* as the official section journal, but there was opposition to this, particularly from Ray G. Stephens (Ohio State University), the 1985-86 section chairman (Stephens also was acting chair during the spring of 1985 due to Ellen Cook being out of the country) [Stephens, 1985, p. 2].

The section membership was surveyed in 1985 concerning the degree of support for a section journal. Approximately 37% of the membership strongly supported a new journal, another 34% was somewhat supportive, 4% were neutral, and 25% were opposed to the section starting a journal. The major reason for support was a lack of outlets for behavioral articles in existing accounting journals. The major reason for opposition was the lack of quality manuscripts to publish in a new journal. With this background, the section officers decided at their November 1985 meeting in Chicago that the section should take a middle road and publish an occasional volume of papers on behavioral issues related to accounting. This decision was reached because of the support indicated in the membership survey. Since the availability of quality papers could not be tested empirically, the officers felt that an interim step would be to publish an occasional volume. It was felt that publication of an occasional volume would allow the section to gain experience in publishing without the full costs of a journal. The first volume was scheduled for mid 1987 ["ABO Section....," 1986, p. 6].

Ken Euske (Naval Postgraduate School) was named in 1986 as the first editor of the new journal (called a collection of papers at that time). A grant of $5,000 was received from the National Association of Accountants to fund the first issue. The grant was the work of Stephen Landekich, a past section vice chairman-practice, who was the NAA research director. No such financial support was received for subsequent issues—thus necessitating a dues increase from $10 to $15 in 1990.

The first issue of what was to become the section's annual journal, *Behavioral Research in Accounting (BRIA)*, was published in 1989—two years later than originally scheduled. The first issue contained eight articles, five of them commissioned by the editor. Those five commissioned articles could essentially be called history articles in that they dealt with the traditions and background of behavioral accounting research. The editor emphasized that the high number of commissioned articles would not be a standard practice, but he felt that the first issue should present a robust history of the field and a critical commentary on its past, present, and future [Euske, 1989, p. ii]. Euske also edited the 1990 and 1991 volumes which relied more heavily on regularly submitted articles. On July 1, 1991, Kenneth R. Ferris (Southern Methodist University) took over as editor and will be responsible for the 1992 edition.

Seminars and Other Meetings

Throughout its history the ABO Section has organized and sponsored seminars on various aspects of behavioral accounting. The first such seminar, organized by vice chairman Eric Flamholtz, was a two-day conference entitled "Human Resource Accounting: The Second Wave," which was held in May 1983. Also in 1983 was a New Orleans symposium entitled "Adapting to a Changing Environment." This latter meeting was co-organized by Tony Tinker and Gerald H. B. Ross.

The ABO Section sponsored seminars on the day prior to three of the 1984 AAA regional meetings. The most comprehensive of these was at the Western Regional held in Tucson, where Ken Euske and Paul Watkins developed a seminar entitled "Current Issues and Developments in the Behavioral Sciences: Implications for Accounting Education and Research." In 1986, Denise Nitterhouse organized a behavioral research methodology program on the day preceding the Midwest Region meeting in Chicago.

In July 1989, the ABO Section cosponsored a research conference with the European Management Control Association at the London Business School. Kenneth Merchant (Harvard University) represented ABO on the organizing committee. The section provided a $500 travel grant to non-European members whose papers were accepted for the conference. These travel grants were designed to broaden participation in the conference. Also, accepted papers were considered for publication in *Behavioral Research in Accounting*.

This listing of seminars is not intended to overlook the many activities at the regional level. In the early years, the regional coordinators did an excellent job of just getting behavioral sessions added to the regional meetings.

Other ABO Section Activities

In 1986, the ABO Section established an outstanding dissertation award. There were 12 submissions the first year, 11 in 1987, 20 in 1988 and 14 in 1989. Initially, there was no money accompanying the award, but in 1988, a $250 prize was added. In 1989, the award structure was changed to give $1,000 to the first place winner and $500 to the runner-up. The first winner of this award was Joseph G. Fisher, Ohio State University, for his 1986 dissertation entitled "The Allocation Mechanism of Audits: An Experimental Approach."

Summary of ABO Section

The ABO Section has grown quickly. There were only 224 members in 1982. By summer 1988, the section had 815 members, including 176 non-U.S. members representing 36 different countries. Today, there are over 1,100 members. The reason for this growth is primarily due to the research interests of its members. The section provides a service by communicating what others in the field are doing. Although the section has sponsored programs and a couple of book-length publications over the years, the primary output has been the semiannual newsletter and recently the annual journal.

Despite a committed group of officers, the productivity of the section has always been slow in developing. Virtually every endeavor has taken longer than was anticipated. Perhaps Frank Collins said it best when he was chairman in 1986-87:

> While in previous years our section has made great strides, there have
> been occasions where we promised more than we produced [Collins,
> 1986, p. 1].

Despite the delays, the membership has not abandoned the section, it is now the fifth largest of the AAA sections. Apparently the products of the ABO Section have been worth waiting for.

AMERICAN TAXATION ASSOCIATION

As mentioned earlier in this chapter, the American Taxation Association (ATA) was one of the groups that led the AAA Executive Committee to move toward the acceptance of sections. The history of the ATA has been well documented in two articles by D. Larry Crumbley in the *Journal of the American Taxation Association* [1987] and *The Accounting Historians Notebook* [1989]. Much of the following paragraphs is taken from Crumbley's articles. Crumbley was the founding father of the ATA and served as its first president in 1974-75. He has since served as the section historian. Other officers are listed in Exhibit 5-7. All officers back to 1974 are listed. However, prior to the 1978-79 fiscal year, the ATA was not an AAA section.

In August 1976, the ATA membership voted to seek AAA sectional status. By then there were over 850 ATA members, far more than the 100 needed to form an AAA section. ATA president Victor Tidwell applied for sectional status on October 15, 1976. The application was contingent upon the ATA being able to keep its separate corporate structure. The AAA Executive Committee accepted this contingency, but indicated that the ATA corporate charter should be given up within three years since by that time it would be known whether the affiliation was flourishing. Also, the Executive Committee objected to an ATA request that its president sit on the AAA Executive Committee [Minutes, November 1976, p. 11]. As a result, the ATA officers declined the opportunity to join AAA as a section.

ATA President Don Skadden submitted another application to the Executive Committee in early 1978. This application was accepted. Subsequently, the ATA membership voted in August 1978 to officially become a section. The ATA corporate charter was terminated in late 1982 following an official merger with AAA.

The objectives of the ATA are four in number:

1. To promote tax education through course design, teaching methods, and classroom aids.
2. To promote dissemination and publication of information on taxation.
3. To promote research on tax policy, tax proposals, and tax legislation.
4. To encourage interaction between various disciplines and professionals through the development of tax-oriented educational and research programs and materials.

Exhibit 5-7
American Taxation Association Officers

Year	Chairman	Chair Elect	Vice Chairman	Secretary	News Editor
1975	D. Larry Crumbley		Ronald N. Bagley	Francis B. Tims	D. Larry Crumbley
1976	Ray M. Sommerfeld	Victor H. Tidwell	Larry C. Phillips	Francis B. Tims	D. Larry Crumbley
1977	Victor H. Tidwell	Donald H. Skadden	Patricia C. Elliott	G. Fred Streuling	Dale L. Davison
1978	Donald H. Skadden	Albert R. Mitchell	N. Allen Ford	G. Fred Streuling	Dale L. Davison
1979	Albert R. Mitchell	Allen Ford	Jane O. Burns	Boyd C. Randall	Kenneth W. Milani
1980	Allen Ford	William L. Raby	Lyle Petit	Boyd C. Randall	Kenneth W. Milani
1981	William L. Raby	Lawrence C. Phillips	John C. Williams	Edward E. Milam	Edward C. Foth
1982	Lawrence C. Phillips	James E. Wheeler	Barry C. Broden	Edward E. Milam	Edward C. Foth
1983	James E. Wheeler	G. Fred Streuling	Robert M. Rosen	Sally M. Jones	Kevin M. Misiewicz
1984	G. Fred Streuling	James H. Boyd	Willis C. Stevenson	Sally M. Jones	Caroline Strobel
1985	James H. Boyd	Jack Kramer	Michael L. Moore	Sally M. Jones	Caroline Strobel
1986	John L. Kramer	Sally M. Jones	John O. Everett	Edmund Outslay	Paul J. Streer
1987	Sally M. Jones	Michael L. Moore	W. E. Seago	Edmund Outslay	Paul J. Streer
1988	Michael L. Moore	Jane O. Burns	Kenneth H. Heller	William N. Kulsrud	Robert L. Gardner
1989	Jane O. Burns	W. Eugene Seago	Sandra S. Kramer	William N. Kulsrud	Robert L. Gardner
1990	W. Eugene Seago	Kenneth H. Heller	Anna C. Fowler	Edward J. Schnee	Barry P. Arlinghaus
1991	Kenneth H. Heller	Edward J. Schnee	Caroline P. Strobel	Debra W. Hopkins	Barry P. Arlinghaus

These objectives are accomplished through the publication of a journal, a newsletter, and special monographs. Tax education is promoted through committee activities, sponsoring awards and educational programs, and through coordination of regional and national meetings. The ATA has surely been the most active of the AAA sections. It has more committees, committee members, programs, and publications than any other section. The ATA was the first section to establish its own journal when *The Journal of the American Taxation Association (JATA)* was started in 1979. The first editor was G. Fred Streuling (Brigham Young University). He was followed in 1981 by John L. Kramer (University of Florida); Jane O. Burns (Indiana University) served from 1984-87 and Sylvia Madeo (University of Missouri at St. Louis) from 1987-90. The current editor is Edmund Outslay (Michigan State University).

The ATA newsletter is published three times per year (spring, summer, fall). The first ten volumes, through summer 1984, consisted of three issues each. However, a numbering problem apparently resulted in volume 11, which included four issues beginning with fall 1984 and ending with fall 1985. Following the fall 1985 issue, every issue has been designated a separate volume. Thus, there are now three volumes per year. For example, the fall 1990 issue was volume 26—a short five years after volume 11.

ATA Awards

The ATA has established an outstanding manuscript award and an outstanding dissertation award. The Tax Manuscript Award was begun in 1982 for an outstanding article published during the preceding three years. The winners of these awards include:

1982   Silvia A. Madeo
1983   Edmund Outslay and James E. Wheeler
1984   Michael W. Maher and Timothy J. Mantell
1985   Charles R. Enis and Darryl L. Craig
1986   Michael L. Moore, Bert M. Steece, and Charles W. Swenson
1987   Karen S. Hreha and Peter S. Silhan
1988   Silvia A. Madeo, Albert A. Schepanski, and Wilfred C. Uecker
1989   Michael L. Moore, Bert M. Steece, and Charles W. Swenson
1990   Paul J. Beck and Woon-Oh Jung

An annual outstanding dissertation award was begun in 1981. These awards, which include a $5,000 stipend, have gone to the following individuals (doctoral schools in parentheses):

1981   Steven T. Limberg (Arizona State University)
1982   Barbara A. Ostrowski (University of Illinois)
1983   Kenneth E. Anderson (Indiana University)
1984   Carol Olson (University of Florida)
1985   Valerie C. Milliron (University of Southern California)
1986   Wayne H. Shaw (University of Texas)
1987   Steven J. Crowell (University of Georgia)
1988   Suzanne Luttman (University of Illinois)
1989   James C. Young (Michigan State University)
1990   Amy E. Dunbar (University of Texas)

Initially, the dissertation award was funded by the CPA firm of Alexander Grant & Company. Touche Ross funded the 1984 award. Since 1985 the award has been funded by Price Waterhouse, except in 1986 when Peat Marwick Mitchell sponsored the award.

Another award sponsored by the ATA is the Ernst & Young Tax Manuscript Award which began in 1989. This award, consisting of a $2,500 stipend, is given to the best manuscript written by a student in a master's program.

Annual Meeting Luncheons

One of the highlights of the AAA annual meetings has become the luncheons sponsored by the ATA. The speakers at these luncheons have typically been the top people in the tax field, resulting in attendance by many non-ATA members. Speakers over the years have included:

1979— Spark M. Matsunaga (U.S. Senator)
1980— Wilbur Mills (Former Congressman)
1981— Roscoe L. Egger (IRS Commissioner)
1982— Bernard Shapiro (Price Waterhouse)
1984— Michael J. White (Bank of Montreal)

1985— Peter J. Panuthos (U.S. Tax Court judge)
1986— Gerald Padwe (Touche Ross & Co.)
1987— Alan Murray (Wall Street Journal reporter)
1988— Norman Ture (former Undersecretary of Treasury)
1989— Spark M. Matsunaga (U.S. Senator)
1990— Fred T. Goldberg, Jr. (IRS Commissioner)

Another ATA activity that is rather new is the mid-year conference held each February since 1989. The first such meeting in Dallas was a one-day program centered around the theme of "Teaching Taxation Courses." The 1990 meeting in Atlanta was extended to two days and covered both teaching and research, as did the 1991 meeting in Albuquerque.

Summary of ATA

As Crumbley pointed out in his 1987 history of the ATA, the group has progressed from orphans to outlaws to respectability. The ATA was at the heart of the AAA sectional movement. Subsequently, it has become the most active and innovative section. For example, in 1990 there are 20 ATA committees, most of which have 15 or more members. The ATA has definitely accomplished its research and education objectives, but it has been somewhat lax in meeting its objective regarding tax policy and the influencing of tax legislation. Still, by getting more educators interested in the field of taxation, the role of the ATA is continually changing, and that change may be toward the influencing of legislation and tax policy. The ATA now has over 1,200 members and is the third largest of the AAA sections. The high level of activities pursued by the ATA should serve as an example to other AAA sections. As 1989-90 ATA president Eugene Seago pointed out in a letter to members:

> While the changes in tax education and practice may appear ominous, the future of the ATA has never looked brighter. This is true because the organization is based on an infallible formula: a well-defined group of competent and self-motivated individuals with common interests. ... Change will merely present new opportunities to accomplish collective goals [Seago, 1990, p. 1].

PUBLIC INTEREST SECTION

Provisional status was granted for a Public Interest Accounting Section at the March 1979 Executive Committee meeting. The objectives of the proposed section were to:

1. Facilitate interaction among AAA members who are interested in accounting as it relates to corporate social responsibility, public issues, and public assistance.

2. Establish programs and develop educational materials aimed at developing an awareness within each accounting student of the accounting professional's

obligation to be concerned with and to bring his/her unique expertise to bear on general societal problems.

It was noted that these objectives would encompass public issues cases, direct accounting assistance to individuals in small business, taxpayer assistance, direct assistance to non-profit organizations, and measurement of corporate performance as it relates to social issues.

The Public Interest Section of AAA was established on March 14, 1980 [Minutes, March 1980, p. 10]. Mark Asman (Bowling Green State University) was the first section chairman. Other officers are listed in Exhibit 5-8.

The Public Interest Section has seen slow growth over its history. Membership in 1990 stood at 334.

Exhibit 5-8
Public Interest Section Officers

Year	Chairman	Vice Chairman	Secretary
1981	Mark Asman	Scott S. Cowen	Paul H. Smith
1982	Scott S. Cowen	Jeffrey Harkins	Paul H. Smith
1983	Jeffrey L. Harkins	Anthony M. Tinker	Paul H. Smith
1984	Anthony M. Tinker	Barbara D. Merino	Paul H. Smith
1985	Barbara D. Merino	Marcos Massoud	Paul H. Smith
1986	Marc Massoud	Cheryl Lehman	Alan Cherry
1987	Marc Massoud	Cheryl Lehman	Alan Cherry
1988	Cheryl R. Lehman	Paul F. Williams	Jean DuPree
1989	Paul F. Williams	C. Edward Arrington	Jean DuPree-Martin
1990	Joanne Rockness	Anthony M. Tinker	Paula B. Thomas
1991	Anthony M. Tinker	Jere R. Francis	Paula B. Thomas

MANAGEMENT ACCOUNTING SECTION

The Management Accounting Section was established in March 1982 when the Executive Committee unanimously approved a request from William L. Ferrara (Pennsylvania State University). Ferrara served as the first chairman of the section. Other section leaders are listed in Exhibit 5-9. Like the Auditing Section, the Management Accounting Section nominating committee develops a slate of two candidates for each position and officers are elected in contested elections.

It is surprising that the Management Accounting Section was so slow in being established. As mentioned earlier, Myron Gordon was advocating such a section in 1965, and once Ferrara got the section going, members were quick to jump on the bandwagon. Over 200 people attended the organizational meeting in San Diego in August 1982. Over 625 had paid their dues by February 1983, and today the membership exceeds 1,600 making Management Accounting the second largest section.

A 28-member coordinating committee was appointed at the 1982 organization meeting to get the section into operation. Interestingly, all of the section's first nine chairmen served on that committee. The committee was divided into six task forces: by-laws, newsletter, meetings, instruction, research, and membership. The coordinating committee held a meeting in Dallas in December 1982 at which time the by-laws were approved for presentation to the membership.

The section newsletter, *Management Accounting News & Views,* appeared in the spring of 1983 under the editorship of Terry Campbell (University of Central Florida). The first newsletter subcommittee decided that the newsletter should be a slick, high quality publication from the start, and this it was. The newsletter has always been published in Sarasota, despite the fact that Shane Moriarity offered to publish the first year's issues at the expense of the University of Oklahoma ["Report of...," 1983, p. 6]. The first newsletter contained news items and minutes of meetings, but beginning with the second issue, in 1984, the newsletter has also contained numerous by-lined articles.

A journal was started in 1990 under the title *Journal of Management Accounting Research.* William Ferrara was the editor of the first two issues and was replaced by Chee Chow (San Diego State University) in 1991. When the journal idea was first approved in 1987, the title was to be *Management Accounting Research,* but complaints from the National Association of Accountants that such a title could be confused with their journal, and the subsequent publication of a British journal with the same name, caused the section to rename its journal.

The first book-length publication of the section was a 1984 annotated bibliography on management accounting by Michael Maher and Charles Klemstine, which was published and distributed to all section members at a cost exceeding $4,800.

In 1983, the National Association of Accountants (NAA) established a Committee on Academic Relations (CAR), and soon thereafter hired a director of academic relations. With the NAA committee and the AAA section being established at about the same time, the result has been an increased posture of management accounting in academia. The two groups have worked together on many projects and seminars over the years, and many individuals have served on both CAR and committees of the Management Accounting Section. The result has been that the section has

---

Exhibit 5-9
Management Accounting Section Officers

Year	Chairman	Vice Chairman	Secretary	News Editor
1983	William L. Ferrara			Terry Campbell
1984	William L. Ferrara	James M. Fremgen	Ronald V. Hartley	Terry Campbell
1985	James M. Fremgen	Thomas Klammer	Ronald V. Hartley	Terry Campbell
1986	Thomas Klammer	Germain Boer	Donald K. Clancy	Shane Moriarity
1987	Germain Boer	Ronald V. Hartley	Donald K. Clancy	Shane Moriarity
1988	Ronald V. Hartley	Shane Moriarity	Terry Campbell	Robert Capettini
1989	Shane Moriarity	Donald K. Clancy	Terry Campbell	Robert Capettini
1990	Donald K. Clancy	Robert Capettini	L. Gayle Rayburn	Lanny Solomon
1991	Robert Capettini	Donald Madden	L. Gayle Rayburn	Lanny Solomon

been able to get increased exposure for many of its programs. One of the most successful of these joint programs has been the Management Accounting Symposia, similar in format to the Trueblood Seminars, which began in 1985.

In summary, the Management Accounting Section has been the fastest growing AAA section, and has profited from its partnership with the NAA Committee on Academic Relations. The only oddity about the section is why it took so long to form.

GENDER ISSUES IN ACCOUNTING SECTION

The AAA's newest section is the Gender Issues in Accounting Section (GIAS), which received final approval from the Executive Committee in August 1990. Ann B. Pushkin (West Virginia University) was the first section chairperson (1989-90). Paula Thomas (Middle Tennessee State University) followed. Cheryl R. Lehman (Hofstra University), the moving force behind the establishment of the section, was the first vice chairperson and director of research. The first secretary-treasurer was Robert A. Nehmer (Ohio State University).

The initial dues of the Gender Section were set at $2 per year—an artificially low amount in order to encourage sufficient membership to meet the AAA's requirements of 400 members to form a section—a ploy that was apparently successful.

References

"ABO Section Occasional Volume," *ABO Newsletter*, Spring, 1986, p. 6.

Anderson, Wilton T., Letter to Arthur G. Mehl, April 6, 1976.

"Auditing Section Establishes New Journal," *Accounting Education News*, January, 1981, p. 12.

"Board of Governors Considers Proposed Standards," *Newsletter: Administrators of Accounting Programs*, Fall, 1979, p. 7.

Brummet, R. Lee, Interview by Terry K. Sheldahl, October, 1989.

Coffman, Edward N., Alfred R. Roberts, and Gary John Previts, "A History of the Academy of Accounting Historians, 1973-1988," *The Accounting Historians Journal*, December, 1989, pp. 155-206.

Collins, Frank, "Memo From the Chair," *ABO Newsletter*, Fall, 1986, p. 1.

Crumbley, D. Larry, "The Evolution of the ATA: From Orphans, to Outlaws, to Respectability," *The Journal of the American Taxation Association*, Fall, 1987, pp. 86-100.

Crumbley, D. Larry, "The Evolution of the American Taxation Association," *The Accounting Historians Notebook*, Spring, 1989, p. 22.

Dittenhofer, Mortimer, "Point of View," *Public Sector Section News*, Summer, 1982, p. 4.

Edwards, James Don, "Comments From the President," July, 1971, pp. 390-392.

Euske, Ken J., "Editorial," *Behavioral Research in Accounting*, 1989, p. ii.

Granof, Michael, "Message from the Chairman," *Public Sector Section News*, Spring, 1982, p. 2.

Holtfreter, Robert E., "Chairman's Report," *ABO Newsletter*, Spring, 1984, pp. 1-2.

Marquette, Penny, *Regional Coordinators' Handbook*, Sarasota: American Accounting Association, 1986.

Mehl, Arthur G., Letter to Administrators of Accounting Programs, June 29, 1976.

Minutes of the American Accounting Association Executive Committee Meetings, 1965-1990.

Minutes of Meetings, Board of Governors: Administrators of Accounting Programs, 1976-1990.

"Report of the Newsletter Subcommittee," *Management Accounting News & Views*, 1983, p. 6.

Rhile, Joe, *The Two-Year College Faculty Directory* (Cincinnati: South-Western Publishing Company), 1989.

Ross, Gerald H. B., "From Your Section Chairman," *Accounting, Behavior & Organizations Newsletter*, May, 1982, p. 1.

Seago, Eugene, "President's Letter," *ATA Newsletter*, Summer, 1990, p. 1.

"Section Annual Meeting," *Government & Nonprofit News*, Fall, 1984, p. 2.

"Section Annual Meeting Elects Officers; Modifies Education, Research Programs; Considers Journal Sponsorship," *Public Sector Section Newsletter*, Fall, 1981, pp. 1-2.

"Section Conducts Successful Meetings in San Diego," *Public Sector Section News*, Fall, 1982, pp. 4-5.

"Sections," *Accounting Education News*, April, 1975, p. 1.

"Sections Organize," *Accounting Education News*, October, 1976, pp. 1-2.

"Sections: Pro or Con?," *Accounting Education News*, December, 1974, p. 1.

Sprouse, Robert, Interview by Dale L. Flesher, December, 1989.

Stephens, Ray G., "Message from the Chair," *ABO Newsletter*, Summer, 1985, pp. 1-2.

"Summary of Minutes—Board of Governors Meeting, February 15, 1979," *Newsletter: Administrators of Accounting Programs*, Spring 1979, p. 2.

CHAPTER 6
EDUCATIONAL PROGRAMS INVOLVEMENT

The Association's first 20 years were devoted primarily to educational matters such as raising the standards for teaching accounting at the collegiate level and building respect for the accounting educator [Zeff, 1966, p. 88]. Over the years, several *ad hoc* committees dealt with course and curriculum content. By 1936, the educational emphasis was replaced by a research emphasis. During 1935 and 1936 several determined members succeeded in reorganizing the Association and changing its name from the American Association of University Instructors in Accounting to the American Accounting Association. The reorganization included a provision in the bylaws for an aggressive program of sponsored research and publications, as well as the addition of a research director to the executive committee. It would be another 33 years before an education director was added to the executive committee.

Despite the emphasis on research and the establishment of accounting principles during the Association's second quarter century, there were many education-related committees and at least one major publication. The committees of that period studied the recommendations of the Ford Foundation and Carnegie reports on business education, discussed whether the Association should participate in the accreditation of accounting programs, explored the opportunities and advantages of faculty residencies, and examined the arguments for professional schools of accountancy [Zeff, 1966, p. 85]. During the early 1950s, an Accounting Careers Committee was formed with the objective of attracting more and better high school students into university accounting programs. With the joint support of the AICPA, FEI, IIA, and NAA, that committee became known as the Accounting Careers Council, and in 1965 was spun off from the AAA as an independent entity. In 1953, the AAA sponsored a 258-page volume entitled *Accounting Teachers' Guide*, which was published by South-Western Publishing Co. The book was an introduction to teaching for young instructors and had been edited by a committee chaired by Leo Schloss of Long Island University [Zeff, 1966, p. 59]. The fellowship program for doctoral students was started in 1958 as the result of a grant from the Haskins and Sells Foundation. In addition, numerous articles of a pedagogical nature were published in "The Accounting Exchange" and "The Teachers' Clinic" sections of *The Accounting Review.*

A Revised Teachers' Guide

The publication of a revised version of *Accounting Teachers' Guide* was one of the first education-related activities of the AAA's third quarter century. During 1965-

66, a committee met to decide whether the 1953 volume should be revised or allowed to go out of print. The decision was to develop a major revision while retaining the basic format of the original book. Although called a second edition, the 1968 volume carried a different title—*A Guide to Accounting Instruction: Concepts and Practices.* The revised version was authored jointly by several members of the 1966-67 Committee to Prepare a Revised Accounting Teachers' Guide. The chairman of the committee was Doyle Z. Williams, then manager of special education projects with the AICPA. Other committee members were Clarence G. Avery (Northern Illinois University), I. E. McNeill (University of Houston), James W. Pattillo (Louisiana State University), A. J. Penz (University of Alabama), William C. Tuthill (Emory University), and Roy E. Tuttle (University of Wisconsin). Penz had also been a member of the committee which had prepared the 1953 volume.

The purpose of the book, which like its predecessor was published by South-Western Publishing Co. of Cincinnati, was to provide reference material which would be useful to both the experienced and inexperienced accounting instructor. The Preface recommended that copies of the book be used in courses aimed at aspiring accounting teachers and that a copy should be given to every graduate teaching assistant at the time of their appointment. The opening chapter, authored by A. J. Penz, was essentially an updated version of the first chapter of the previous edition and outlined the history of accounting education and the professional standards and ethics of accounting professors. The basis of the professional standards section was adapted from a publication of the American Association of University Professors (AAUP) [Committee to Prepare..., 1968, pp. 1-24].

A New Emphasis on Education

In 1966, the Association's Long-Range Planning Committee made several recommendations concerning the AAA role in accounting education. One proposal was for an educational program similar to the then recently redesigned research program. Other ideas included the publication of education monographs, a project to examine suitable topics for educational research, and the creation of the position of Director of Education who would sit on the Executive Committee. As a result of these suggestions, the Executive Committee, in 1966, established an Educational Research Project, headed by Lyle E. Jacobsen, to investigate educational research, identify problems most suitable for basic research, to describe a role for the AAA in educational research, and to suggest measures for how that role should be implemented. The committee's report offered little in the way of startling recommendations and was never acted upon by the Executive Committee [Zeff, 1970, p. 4].

The 1966-67 Long-Range Planning Committee also recommended an Executive Committee member be added for education—in this instance, a vice president of education. As was true of its predecessor, the 1966-67 Long-Range Planning Committee also recommended a standing committee on education. Subsequently, a subcommittee of the Executive Committee meeting during 1967-68 drew up new bylaws which included a director of education [Minutes, March, 1968, p. 12]. The membership approved the new bylaws at the San Diego convention in 1968. The position of Director of Education-Elect was approved in a 1983 bylaws change.

Directors of Education

The first director of education was Stephen Zeff of Tulane University. As the first incumbent in the position, Zeff had the opportunity to define the position. After serving in the position for one year, Zeff submitted a report to the Executive Committee which reviewed the educational objectives of the AAA, endeavored to translate those objectives into activities, and suggested means by which the activities could best be carried out [Zeff, 1970, p. 1]. He concluded that all of this discussion essentially outlined the role of the director of education, as he saw that role. Although Zeff had specific proposals for the role of the Director, the overall tone of his report was "that the Director of Education should be given the time and funds to stimulate innovation and change in the educational sphere" [Zeff, 1970, p. 9].

Zeff's specific proposals for AAA involvement in educational activities were five in number. They were:

1. Provide continuing education for accounting professors.
2. Maintain a dialogue on curricula, educational philosophy, and teaching technique.
3. Stimulate experimentation with new approaches to teaching.
4. Encourage and support research on the effectiveness of approaches to teaching and of different curriculum patterns.
5. Improve accessibility of teaching resources [Zeff, 1970, pp. 6-9].

The last of the above suggestions involved the AAA reprinting classic accounting books and preparing videotapes of well-known accounting personalities. The Association never got involved in any aspect of this recommendation (other than a 1983 videotape of recipients of the Outstanding Educator Award). In the case of books, private enterprise, in the form of Scholars Publishing Company, Arno Press, and later Garland Press, filled the void. Similarly, Michigan State University, the Touche Ross Foundation, and the Academy of Accounting Historians filled the void with respect to videotaping of accounting personalities.

The power of the Director of Education was formally increased at the August 1970 Executive Committee meeting when approval was given for the Director to have explicit authority to authorize Association publications up to the limit of his or her budget [p. 7]. In other words, the Director does not have to have the formal approval of the Executive Committee to accept something for publication.

Zeff's first year was partly devoted to the area of professorial development, "an area of Association activity that had become tangled in misunderstandings between the Executive Committee and the Committee on Professorial Development regarding the objectives, the publication of course materials, expenditure policies, and, in general, the allocation of authority" [Zeff, 1970, p. 5].

The Committee on Professorial Development began developing continuing education courses for professors in 1965 at the urging of 1966 president Herbert E. Miller. Dennis E. Grawoig, head of the Quantitative Methods Department at Georgia State University, was the initial chairman of the committee. The first course, developed and taught by Grawoig, was a two-week program on quantitative methods that was offered for the first time just prior to the 1967 annual meeting. A second course followed the next year in the form of a one-week program on behavioral sciences and accounting. The courses were offered in the same locale as the annual convention. These courses were developed and taught by members of the committee. The committee members were being paid to teach these courses—a factor that resulted in criticism from the Executive Committee because it was felt that no AAA member should receive pay for serving on a committee—and that was essentially what was occurring. In April, 1970, a new policy was adopted that gave the Director of Education the authority over programs after they had been developed. The Professorial Development Committee had charge of the planning and development activities. Zeff's first report to the Executive Committee requested that the Director be released from the burden of overseeing on-going professorial development courses, and that this responsibility be assigned to a new director of professorial development [Zeff, 1970, p. 12].

Initially, the Director of Education had no budget, but in December, 1969, the president allocated $1,000 for travel costs. During the first year, Zeff attended all meetings of the Committee on Professorial Development, and advised that committee on the preparation of a document clarifying the committee's authority in the development of new courses. In addition, Zeff authorized an "Education Series" of occasional papers for the publication of the outlines and lecture notes of the professorial development courses. The first volume, published in mid 1971 and funded by the Ernst & Ernst Foundation, was entitled *Lecture and Problem Materials in Quantitative Methods* and was coauthored by Charles L. Bostwick, Dennis E. Grawoig, Edward G. Rodgers, and William J. Thompson, all of Georgia State University. This volume consisted of the problems, with solutions, that were used in the first professorial development course. Although the book could be used for self study, it was of little value other than as a textbook for the course. The book was designed to "stimulate accounting professors to enhance their understanding of the increasing number of links between quantitative methods and accounting" [Bostwick, 1971, p. v].

Zeff received two research proposals during his first year in office, one of which was funded. That project dealt with the development of a time-sharing computer program library for use in accounting curricula.

Zeff also obtained funds from the Touche Ross Foundation for a Symposium on Behavioral Science Research in Accounting. Technically, no member of the Executive Committee except the president has the authority to request funds from outside organizations. Knowing this, Zeff went to Robert Trueblood at Touche Ross and described his proposal for the Behavioral Science Symposium. He then explained that he was not authorized to request funding for the program; however, if Touche Ross were to offer funding for a program such as he had described, then he would be more than happy to take the offer to the Executive Committee. The offer was made, as Zeff had hoped, and the Executive Committee was quick to accept. The papers from the October 1970 Symposium were never published because they were presented in the form of working papers, but Zeff has a complete set, of which he is quite proud [Zeff, 1989].

Zeff's main disappointment during his term as Director of Education was that he was unable to convince the Executive Committee of the need to approve the 1940 Statement of Principles of the American Association of University Professors (AAUP). The 1969 Executive Committee defeated the motion by a 3-to-4 vote, while the 1970-71 committee voted 5-to-5, resulting in non passage [Minutes, August, 1970(a), p. 11, and 1970(b), p. 4].

Harold Langenderfer of the University of North Carolina served as the second director of education from 1971-73. Most of Langenderfer's first year was spent administering the professorial development program, and particularly the need to develop an errata sheet for the quantitative methods book. Ultimately, it was decided that the book contained so many errors that the AAA should not be associated with the volume. The Executive Committee voted at its April 1972 meeting to discontinue sales and distribute remaining copies free to the academic members [Minutes, April, 1972, p. 12]. At the August 1972 meeting, Langenderfer proposed that:

> All copies of *Lectures & Problem Materials in Quantitative Methods* are to be destroyed prior to August 31, 1972. This action is to supercede the decision to send free copies to academic members taken by the Executive Committee at its meeting in April, 1972 [Minutes, August, 1972, p. 6].

This motion was passed unanimously despite the fact that the authors had offered to supply an errata sheet free of charge.

Perhaps the major accomplishment of Langenderfer's tenure as director of education was the establishment of the Outstanding Accounting Educator Award. The Executive Committee approved the award in April 1972. A later section discusses the award in more detail.

Another of Langenderfer's undertakings was the commissioning of James Don Edwards to edit a book entitled *Accounting Education: Problems and Prospects.* There were 132 manuscripts submitted for possible publication in this classic volume.

Langenderfer's final project as director was to write and submit a proposal to the Price Waterhouse Foundation for a symposium to study the key areas in accounting education which needed to be researched. That proposal was subsequently funded in the amount of $100,000.

Doyle Z. Williams, then at Texas Tech University, was the education director from 1973-75. It was under his administration that Volume 1 in the "Accounting Education Series," *Accounting Education: Problems and Prospects*, was published in 1974. This 602-page, hard bound, book was edited by James Don Edwards and represented the work of more than 80 authors. The book included papers on the challenges to accounting education, conceptual framework for education, motivation, communication and learning theory, instructional innovation, evaluation of performance of both students and teachers, and research methodologies applicable in accounting education.

Robert Grinaker, University of Houston, served as the director of education from 1975 to 1977. In 1976, Grinaker published Volume No. 2 in the "Accounting Education Series" entitled *Researching the Accounting Curriculum: Strategies for Change*, edited by William Ferrara (Pennsylvania State University). This volume consisted of papers and critiques presented at a symposium sponsored by the AAA with the support of the Price Waterhouse Foundation. The symposium was designed to identify researchable problems in accounting education and was the first step in furthering the purposes of a Price Waterhouse Foundation Research Grant to the AAA.

Leon Hay, then at Indiana University in Bloomington, was the director of education from 1977 to 1979. In 1979, he oversaw the publication of three research projects including *The Core of the Curriculum for Accounting Majors*, by Richard E. Flaherty (Volume No. 3 in the Accounting Education Series), which was the culmination of a six-year project supported by the Price Waterhouse Foundation. Also published in 1979 were *Accounting Students and Abstract Reasoning: An Exploratory Study*, by George E. Shute, and *Empirical Research in Accounting: A Methodological Viewpoint*, by Rashad Abdel-khalik and Bipin Ajinkya (Volume No. 4 in the Accounting Education Series). The latter study, the outgrowth of an AAA continuing education course entitled "Research Appreciation," was funded jointly from the budgets of the directors of education and research. Hay was also instrumental in obtaining a grant from the Coopers & Lybrand Foundation to stimulate research into innovative methods of teaching accounting. Other activities which Hay administered included the 1979 Community College Workshop at the University of Massachusetts, the production of AAA Videotape No. 3 entitled "Management's Responsibilities for Internal Control: The Impact of the Foreign Corrupt Practices Act," the Trueblood seminars, and the doctoral consortium ["Interim...", 1980, p. 3].

James Fremgen of the Naval Postgraduate School was the education director from 1979 to 1981. He oversaw the publication of Volume No. 5 in the Accounting

Education Series entitled *Developing Communications Skills for the Accounting Profession*, by Robert W. Ingram and Charles R. Frazier. Fremgen was fortunate in having available a $100,000 grant from the Coopers & Lybrand Foundation for the purpose of funding research into instructional methods in accounting. The first publication resulting from the C&L grant was the 1980 *A Review of Innovative Approaches to College Teaching* by Beatrice and Ronald Gross. Fremgen also presided over ten continuing education courses offered at the Boston meeting, plus others at regional meetings and two community college workshops ["Annual Report...," 1980, p. 2].

Thomas J. Burns of Ohio State University served as director of education from 1981 to 1983. Burns was responsible for organizing the first new faculty consortium and the first doctoral programs conference, the publication of the first volume of *Issues in Accounting Education*, the creation of the outstanding educators videotape, and the publication of the outstanding educators booklet. In other words, Burns was an innovator; he looked for unmet needs and tried to create programs to meet those needs.

Donald Madden, University of Kentucky, was the 1983-85 director of education. In 1984, Madden published Volume No. 6 in the Accounting Education Series entitled *Accounting Structured in APL*, by Yuji Ijiri.

Loren A. Nikolai of the University of Missouri was the director of education from 1985-87. Nikolai published Volume No. 7 in the Accounting Education Series, *Computer Assisted Analytical Review System*, by William R. Kinney, Gerald L. Salamon, and Wilfred C. Uecker. This book and accompanying diskette was designed to help auditing students understand the application of analytical review procedures.

Gary John Previts, Case Western Reserve University, served as the director of education from 1987 to 1989. He served as director of education-elect the preceding year, during which he requested the Association sponsor a gathering of former directors of education (Zeff, Langenderfer, Madden, Burns, Fremgen, Hay, and Nikolai attended) to assist him in evaluating and mapping future education initiatives for the AAA. During that session, the progression from the Doctoral Consortium to the New Faculty Consortium was discussed as the initial steps in a life-long stair-step series of consortia which perhaps would eventually include both a newly-tenured faculty consortium and a senior faculty consortium. The latter soon came to fruition. During his term, Previts oversaw the publication of Volume No. 9 in the Education Series entitled *A Framework for the Development of Accounting Education Research*, by Jan Williams, Mikel Tiller, Hartwell Herring, and James Scheiner. The book was subsidized by an $8,500 grant from the Coopers & Lybrand Foundation. The study itself was also funded by the C & L Foundation. This study addressed the methodology and subject areas of accounting research in the areas of students, faculty, and the process of education. The authors had been selected by Tom Burns while he was director of education.

Corine T. Norgaard (University of Connecticut) was the first woman to be elected director of education, serving from 1989 to 1991. She presided over the first Senior Faculty Consortium which was offered in June 1990. She also worked with the Education Advisory Committee in its work with the Accounting Education Change Commission in developing a symposium on models of accounting education. Norgaard also served as an ex officio member of the Accounting Education Change Commission.

Carter Scholarships

Arthur H. Carter scholarships are given each year to students who have completed at least two years of study toward a degree in accounting. Each university in the United States is allowed to recommend at least one student for the award (nominations are limited to one percent of a school's preceding year's graduates). These scholarships are provided by a trust fund of about $650,000 which was established under the will of Mrs. Arthur H. Carter, who was the daughter of Elijah Watt Sells, a pioneer of the profession in the United States, and the husband of Arthur H. Carter, who was managing partner of Haskins & Sells from 1930 to 1941, when he resigned to serve in the War Department during World War II. Mrs. Carter died on October 26, 1973. Initially, there was some question about whether the AAA would agree to participate in the program because Mrs. Carter's will stated that the scholarships were to go to "men." The AAA Executive Committee refused to participate if women were not eligible for the scholarships. This position resulted in a problem for the trustees of the estate because Mrs. Carter's will had been written with the provision that the Association would be the organization administering the program. Consequently, the trustees of Mrs. Carter's estate decided that they could interpret the word "men" as "mankind" and thereby allow women to receive Arthur H. Carter Scholarships [Minutes, March, 1976, p. 2].

The scholarship recipients are selected by an AAA committee. The 1988 awards of $2,500 each were presented to 50 recipients. In 1989, there were 45 scholarships awarded of $3,000 each. A total of 175 students applied for the scholarships in 1989 ["Arthur...", 1989, p. 21]. In 1990, there were 201 applicants for the 50 scholarships awarded.

Doctoral Fellowships

The AAA's doctoral fellowship program was initiated in 1958 when a grant for the purpose was received from the Haskins & Sells Foundation. Actually, planning had begun as early as the spring of 1957. The shortage of qualified accounting teachers had become apparent during the 1950s and studies had indicated that increasing enrollments at the undergraduate level were not being matched by a correspondingly larger number of doctoral graduates. The initial objective was to give doctoral candidates a financial boost at or near the dissertation stage. During its first decade, the fellowship program funded 142 doctoral candidates—the total outlay amounting to $227,000 [Zeff, 1970, p. 2]. Interestingly, well over 20 percent of this amount was paid out in fiscal year 1970, a year that ended with $57,576 in the fund unspent ["American...," 1971, p. 181]. In more recent years the objective of the fellowship program has changed to one of providing an incentive to get students to enter a doctoral program, rather than waiting until they were in the dissertation stage to provide money. It was felt that the doctoral-granting universities could provide funds to their third and fourth year people and the role of the AAA should be to encourage people to enter the teaching profession. The average yearly grant to students has been about $1,000 every year during the program's history. Given the inflation that has occurred during this time, it might be asked whether today's grants are as meaningful as those of earlier years.

The fellowship program has not been particularly controversial over the years, except perhaps to Charles T. Zlatkovich who was not supportive of the idea of attracting new doctoral students into the profession. At the December 1970 Executive Committee meeting, during a discussion of the future role and structure of the Association, President-elect Zlatkovich suggested that the fellowship money could possibly be used in a better way. He felt that "the Association should not continue to encourage more Ph.D.s in accounting now that the supply is starting to catch up with the demand" [Minutes, December 1970, p. 14]. Zlatkovich pointed out at the December 1971 Executive Committee that, based upon the findings of the AAA Future Professorial Supply and Demand Committee, the Association should not continue to encourage any more people to enter accounting education. Eldon Hendriksen supported Zlatkovich, but recommended that fellowships be limited to individuals planning to teach at junior colleges [Minutes, December 1971, pp. 12-13]. Zlatkovich noted again at the March 1973 Executive Committee meeting that the demand for Ph.Ds in accounting was going to be considerably reduced in the future and hoped that the Executive Committee would give attention to this problem [Minutes, March, 1973, p. 8]. Could it be that Zlatkovich and the aforementioned committee had some erroneous assumptions included in their data?

A 1980-81 committee chaired by Clyde Stickney (Dartmouth) questioned whether the fellowship program could have been an effective mechanism for increasing the supply of qualified teachers of accountancy, given the relatively small dollar amount of each fellowship grant. The committee concluded that "the size of the grant would not seem to have been large enough to induce individuals to pursue doctoral studies in accounting." Even those who accepted grants and did go into doctoral programs did not always finish those programs. According to the aforementioned committee, of those individuals who received grants between 1959 and 1976, only 81.1 percent had completed their degrees by 1980. Only 82.6 percent of grant recipients were still in academe in 1980. For the most part, those who left teaching went with large public accounting firms [American Accounting Association, 1981]

Doctoral Consortia

The AAA Doctoral Consortium was suggested by president-elect James Don Edwards at the April 1970 Executive Committee meeting. Edwards proposed an annual seminar similar to those sponsored by the American Marketing Association. Doctoral students, one from each doctoral granting university, would have their expenses paid by the AAA from a grant provided by the Haskins & Sells Foundation. Initially, the Foundation agreed to donate $15,000 for the 1971 consortium, but that amount was later increased to $18,000 when it appeared that expenses would be higher than anticipated. The first consortium, held in 1971 in Lexington, KY during the week preceding the annual meeting, was headed by Kermit Larson, then of Tulane University. Forty-four students attended [Minutes, August 21 & 22, 1971, p. 12]. Consortium costs came to $15,239.91. In addition, the planning committee spent an additional $2,120, bringing the total close to the amount donated [Minutes, December, 1971, p. 12]. In subsequent years it became necessary to assess participating schools with a registration fee (initially $100) to help cover the consortium costs. Additionally, the consortium lasted five days in 1971 and 1972, but was reduced to four days in subsequent years [Minutes, December, 1972, p. 13].

The primary objective of the Consortium is to improve the quality of future accounting education and research by enriching the experience of outstanding doctoral candidates, who are selected from different universities throughout the United States, Canada, and Europe. The consortium is designed to provide a forum for stimulating a participant to pursue research in significant subject areas. Candidates are able to interact with recognized leaders in accounting education and research. The programs encompass formal presentations of research papers, less structured discussion sessions, and informal meetings of the Consortium fellows and the faculty, these latter gatherings usually occurring around the pool, over dinner, or in the Consortium suite. Participants are doctoral students who have completed at least one year of residency in their Ph.D. programs and have at least one year of residency remaining. The selection of faculty members is supposed to provide a good geographic balance in terms of both school of current affiliation and school of doctoral degree. Although there is no prohibition against individuals repeating as faculty members, AAA policies stipulate that an attempt should be made to bring in a preponderance of new persons each year [Minutes, March 1975, Appendix A].

In the early years, the consortia were always held in the vicinity of the annual meeting during the week preceding the meeting. This allowed doctoral students to attend the annual meeting without having to incur additional travel expenses. When Nicholas Dopuch took over as chairman of the Consortium Planning Committee for the 1985-86 year, he recommended that the date be changed. Dopuch felt that faculty members were always too tired following the Consortium to fully appreciate the annual meeting. Also, he worried that students were being selected because of their desire to interview for jobs at the annual meeting. Dopuch felt that the Consortium was primarily for individuals who had not yet reached the dissertation stage, thus such persons were not yet ready for the job market. Moving the Consortium to a date remote from the annual meeting would allow schools to send their best people without regard to a desire to support the student's job recruitment. Also, since the Consortium would not have to be held in the same locale as the annual meeting, the proposal would allow the AAA to contract with a hotel on a long-term basis and obtain more favorable room rates [Dopuch, 1985]. The proposal was subsequently adopted. In 1988, the Consortium switched to a June meeting time and a permanent location at Lake Tahoe.

During 1980-81, a committee chaired by Clyde P. Stickney (Dartmouth) studied the success of the Doctoral Consortium during its first ten years. The conclusion was that the consortium program had been an overwhelming success in achieving its objectives and that no substantial changes in objectives, structure, or content were needed. Because of this success, there has been little change in the format of the program over the years. The only noticeable difference between the consortia of the early 1970s and the late 1980s was the increased participation of woman. Whereas there was only one woman who attended during the first three years, the number had increased to 30 for the 1990 Consortium. One surprising aspect uncovered by the 1980-81 Evaluation Committee was that over 30% of those who attended the first two consortia were no longer in academia in 1980. Nearly 20 percent of those attending the consortia between 1973 and 1976 were no longer in academia in 1980. Of course, some might argue that since the brightest and best are supposedly selected for the consortia, those same individuals are in demand by public accounting firms. Also, the committee found that only about 78 percent of those who attended the first six consortia had completed their degrees by 1980.

These completion statistics had improved somewhat by the time that another study was conducted by the Doctoral Consortium Planning Committee in 1987. The 1987 study found that through the first 16 years of the program the percentage of graduates was 67.3 percent, with an additional 18.9 percent still in their doctoral programs. Thus, the potential graduation rate would have been 86.2 percent. Only 11.5 percent of the participants had left academia. On average, participants graduated 2.6 years after attending the Consortium.

The Doctoral Consortium program can perhaps best be summed up by a few words from a letter that Yuji Ijiri wrote while Association president in 1983. In writing to Robert L. Steele of the Deloitte, Haskins & Sells Foundation, Ijiri stated:

> Many members of the Association have remarked that the Doctoral Consortium is the best project that the Association has undertaken. The impact of the Consortium on research and on the academic quality of young accounting professors has been truly significant. In addition, the comradeship created by the Consortium among its several hundred "alumni" is a lifetime asset for these people [Ijiri, 1983].

Indeed, the Doctoral Consortium has been all that Ijiri said, and more than James Don Edwards ever dreamed.

Trueblood Seminars

During 1990, Deloitte & Touche celebrated the 25th anniversary of the Trueblood Seminars. However, the AAA has not been involved throughout the entire history of the program. Initially, the seminars for professors were an in-house Touche Ross project headed by partners Robert Trueblood and Henry Korff. Trueblood was an inspirational leader in those seminars, and when he died the firm decided to honor him by naming the seminars for him. Trueblood had been the firm's chairman of the board and had served as president of the AICPA. When the name was changed in 1974, some of the administrative aspects of the seminars were turned over to the AAA along with a $250,000 commitment from the Touche Ross Foundation to finance the programs for three years.

The Trueblood Seminars, of which three are usually held each year, were designed to advance accounting theory and practice and to further accounting education by providing an opportunity for the study of the application of accounting principles and reporting concepts. The seminars are administered by a committee consisting of four Touche Ross representatives and three AAA representatives. The initial members of the AAA Trueblood Committee included Henry C. Korff, Robert J. Sack, Lewis A. Werbaneth, Jr., and Gerald A. Polansky of Touche Ross, and AAA members Elba F. Baskin (then staff director of continuing education), Robert May (University of Washington), and Arthur G. Mehl (Bradley University). A maximum of 30 participants were to be invited to each of the three annual seminars ["Trueblood," 1974, p. 1].

The first seminars administered under the auspices of AAA were held in 1975. Initially, about 25 cases, half of which were auditing and half accounting, were studied by the participants. The number of cases was reduced to 16 in 1983. In 1986 and later years, Touche Ross partners demonstrated the use of the NAARS and other computerized data bases and gave participants the opportunity to conduct their own NAARS searches during the evening hours [Benjamin, 1986, p. 4]. This was a popular feature for those professors who did not have access to the data bases at their own schools. By 1985, the cost was $120,000 per year. Additionally, the time of top partners was also donated ["The Trueblood...," 1983, pp. 6-7].

Touche Ross also sponsors a breakfast for Trueblood Seminar Alumni at each AAA annual meeting. This event was first held in New Orleans in 1983. The breakfast draws about 300 individuals each year. Until recently, the seminars were held in Chicago. Beginning in 1990, the meeting site was moved to the Deloitte & Touche Development Center in Scottsdale, Arizona. Total costs of the three 1990 seminars exceeded $204,000 ["1990 Trueblood...," 1990, p. 25].

Outstanding Accounting Educator Awards

The AAA Outstanding Educator Awards were approved by the Executive Committee in 1972 and first given in 1973. The first recipients were Wilton T. Anderson of Oklahoma State University and Charles T. Horngren of Stanford University. Both men later became AAA presidents. Since then, one or two individuals have been selected each year for the award by a committee appointed by the Association president. The 1973 award winners were not notified in advance of the annual meeting that they were to be honored. In fact, Charles Horngren was hoping to catch an earlier flight home that would have required him to miss the award ceremony. Fortunately, the president contacted Chuck early in the morning of the day the award was to be given and notified him to be available for the award. Horngren stated that being one of the first recipients of the award was the greatest honor that he could imagine [Horngren, 1990]. A complete list of Award winners is shown in Exhibit 6-1.

To insure complete independence by the selection committee, no member of the Executive Committee or nominee for office is eligible to receive the award while serving on the Executive Committee or while a nominee. Also, as a matter of policy, Executive Committee members and nominees refrain from writing letters in support of particular individuals. The Outstanding Educator Award is truly a prestigious award given to accounting educators who have attained the highest respect from their peers. The general criteria for the award include having had influence on students, an impact on the profession, a record of professional and educational innovation, publications, speeches, and other awards and credentials.

During the early years, members were urged to wage organized campaigns for those individuals they felt were deserving of the award. Thus, it was common for the selection committee to receive hundreds of letters supporting a particular candidate. Members making a nomination were encouraged to get as many letters as possible in support of their favorate candidate. However, that approach was soon abandoned in favor of a more professional manner that still encourages nominations, but without the campaigning overtones.

In 1983, Director of Education Thomas Burns (Ohio State University) arranged for a videotape to be made under the sponsorship of Arthur Andersen & Co. of the 15 members who had received the Outstanding Educator Awards. The taping was conducted at the Arthur Andersen Training Center in St. Charles, Illinois. Nine hours of tape were shot, from which an hour-long program was created. The tape was first shown at the annual meeting in New Orleans as one of the luncheon programs. The tape featured the many dimensions of an accounting professor's career. The objective was to provide a film for classroom use which might help attract students into academic careers in accounting ["Outstanding Educators...," 1982, p. 13]. A 32-page booklet entitled *Outstanding Educators: Fourteen Profiles*, accompanied the videotape.

EXHIBIT 6-1
RECIPIENTS OF AAA OUTSTANDING EDUCATOR AWARD

1973	Wilton T. Anderson, Oklahoma State University
1973	Charles T. Horngren, Stanford University
1974	Kenneth W. Perry, University of Illinois
1975	James Don Edwards, University of Georgia
1975	Carl L. Nelson, Columbia University
1976	Sidney Davidson, University of Chicago
1976	Jack J. Kempner, University of Montana
1977	Herbert E. Miller, University of Georgia
1978	Charles T. Zlatkovich, University of Texas
1979	George H. Sorter, New York University
1980	Norton M. Bedford, University of Illinois
1980	David Solomons, University of Pennsylvania
1981	Nicholas Dopuch, University of Chicago
1981	Catherine E. Miles, Georgia State University
1982	Gerhard G. Mueller, University of Washington
1983	Carl Thomas Devine, Florida State University
1983	William A. Paton, Sr., University of Michigan
1984	Robert K. Mautz, University of Michigan
1984	William J. Vatter, University of California at Berkeley
1985	Maurice Moonitz, University of California at Berkeley
1985	Glenn A. Welsch, University of Texas
1986	Yuji Ijiri, Carnegie Mellon University
1986	Joel Demski, Yale University
1987	Thomas R. Dyckman, Cornell University
1988	Stephen A. Zeff, Rice University
1988	Robert S. Kaplan, Harvard and Carnegie Mellon
1989	Robert N. Anthony, Harvard University
1989	William R. Kinney, Jr., University of Texas
1990	William H. Beaver, Stanford University
1990	William W. Cooper, University of Texas

Innovation in Accounting Education Award

A new award for innovation in accounting education was given for the first time in 1990. The intent of the award is to recognize a significant activity, event, or set of materials that are of an innovative nature. Award criteria include innovation, educational benefits, and transferability to other educational institutions or situations. Nominations may include a set of teaching materials, an overall curriculum or program, an organizational change, a creative instructional strategy, educational research, or an insightful teaching approach. The innovation must have been implemented so that evidence of its success can be evaluated.

The first award was presented to Robin Cooper (Harvard) for "Cases in Cost System Design." These cases innovatively introduce students to a new way of thinking about management accounting systems. The materials highlight decisions faced by company managers in actual organizations. The cases introduce new concepts such as activity-based costing and computer-based models ["Award...," 1990, p. 17].

New Faculty Consortium

The AAA sponsored an experimental conference for 52 new faculty members in February, 1983, at the Arthur Andersen Training Facilities in St. Charles, Illinois. The firm also funded the conference. The conference was designed to provide career planning for new doctorates. The intent was to emphasize the teaching, research, and service activities that are expected of faculty members. Eligible participants were those who had finished their doctorate during the preceding calendar year and taken a job at an AACSB-accredited institution. The resident faculty for the consortium consisted of eight associate professors who had been promoted at the university of their first appointment. In addition, 12 of the Association's Outstanding Educator Award winners were in attendance all or part of the week (they were in St. Charles for the taping of the Outstanding Educator's program). Thomas J. Burns, the AAA's Director of Education was the organizer and chairman of the conference ["New Faculty..., 1982, p. 11]. With the exception of the availability of the outstanding educator award winners, consortium audiences in subsequent years were treated to the same programs as those of the first year. These included an "Effective Presentation" session using Arthur Andersen & Co. video equipment, a presentation by Cornell psychology professor James Maas on "Improving Teaching Effectiveness," and a balance of topics involving both teaching and research effectiveness. Arthur Andersen & Co. has continued to sponsor the programs [Madden, 1985, pp. 6-7].

Doctoral Programs Conference

The AAA's first doctoral program conference was held in June, 1983, at Ohio State University in Columbus, Ohio. It was another of Director of Education Thomas Burns conference innovations. The targeted audience is accounting faculty who are involved in doctoral education. Seventy-six representatives from 53 accounting doctoral programs participated in the first conference. During the conference, participants heard 12 position papers on a variety of program issues. Topics included admissions, mentoring, differentiated programs, funding, dissertations, scholarships, ABDs, statistical trends, student expectations, minorities, and various subject fields such as auditing and tax [Doctoral Programs..., 1983, p. 1]. The objective of the conferences is to improve the quality of doctoral education in accounting through an exchange of ideas related to the dimensions of doctoral programs.

Seminars on Professionalism and Ethics

In response to the Treadway Commission Report on fraudulent financial reporting, the AAA Executive Committee in 1987, through president William Beaver's initiative, agreed to develop a program on professionalism and ethics. Beaver's persuasiveness convinced a group of sponsors from both public accounting and industry to commit to support a series of ethics seminars related to teaching ethics in accounting courses. Harold Q. Langenderfer (University of North Carolina) chaired the committee which organized the seminars. The first seminar was held in Montvale, New Jersey in May 1989. The second seminar was held in Atlanta during May 1990. Approximately 70 individuals, from both academia and sponsoring groups, attended each year. Besides Langenderfer, other committee members who were active in both the development and presentation of the programs were Don Baker (Arthur Andersen & Co.), Joanne Rockness (University of North Carolina), Fred Neumann (University of Illinois), and Robert Sack (University of Virginia) ["1990 Seminar...," 1990, pp. 15-17].

Senior Faculty Consortium

The concept for a senior faculty consortium can be traced to a meeting of former directors of education which was held in Houston in February 1987 at the request of director of education-elect Gary Previts. During that session, it was pointed out that the Association had many programs specifically for new faculty, but nothing specifically for older faculty members. Subsequently, President-elect John Simmons, acting upon the recommendation of a two-year task force chaired by Barbara Merino, approached Bernard Milano of KPMG Peat Marwick to ascertain the firm's interest in supporting such a project. The firm was indeed interested and committed to the project. Loren A. Nikolai, a former director of education, was selected to plan and conduct the first senior faculty consortium which was held in Scottsdale, Arizona in June 1990 [Previts, 1990].

Some 75 speakers, discussants, and participants attended the first seminar at the Scottsdale Conference Center. The objective of the Consortium was to provide a forum for senior faculty to share and exchange insights on trends and innovations in teaching and research relevant to accounting education. Each session began with a speaker and two discussants, after which the participants broke into small groups for purposes of discussing the issues raised as they applied to several functional areas. After the small group discussions, the entire group reconvened and a spokesperson for each group summarized the group's deliberations.

Corporate Accounting Policy Seminar

Under the urging of Eugene Flegm (General Motors Corporation), the AAA inaugurated a conference on corporate financial reporting, from the viewpoint of the preparer, in October 1990. The Association's director of education, Gary Previts was very supportive of the concept and worked with Flegm to develop a mission statement that could be used in obtaining funding and other support from the corporate financial community. That mission statement perhaps best describes the purpose of the seminar:

> The purpose of the seminar is to create an environment whereby those involved in teaching financial accounting and those involved in the non-public practice of accounting can gain appreciation of the problems involved in the application of accounting theory in a business environment.

The conference was by invitation only; 60 academics and 30 corporate financial leaders were invited. James Don Edwards was the first conference coordinator. Flegm had been responsible for raising the funds to cover the costs of those attending the conference. Twenty-nine corporations provided funds and participants as did the Financial Executives Institute, the National Association of Accountants, and the Academy of Accounting Historians.

Discussions relating to the concept of the seminar were initiated as early as 1987. Flegm, Previts, and several corporate financial officers met in New York in June 1988 to assess the interest of the corporate community. The AAA Executive Committee endorsed the concept for the seminar in 1989 after Flegm had obtained promises of financial support from the corporate financial community.

The first seminar, held in New Orleans, was generally acknowledged by participants as an outstanding program, perhaps because corporations gave not only money, but the time of their top people as well. Professors were not only exposed to accounting ideas from a preparer's viewpoint, but were able to rub shoulders with the most powerful financial men and women in America at the same time. Speakers at the seminar included Eugene Flegm and William Ihlenfeldt (Shell Oil).

Summary and Conclusions

In many respects the past quarter century could be termed a building-block period for the AAA. At the beginning, the Association did not even have a Director of

Education. Once the directorship was created, new educational programs began appearing nearly every year. And, once a program was established, it has continued to the present day. Thus, the Association's current educational package has been built up over a quarter century. Does this mean that the Association has never had a failing educational program, or does it mean that programs, once started, are never evaluated? Whatever the case, as the years go by, the job of the Director of Education could become more and more of a caretaker responsibility leaving little time for innovation.

The past quarter century has been a dynamic one in the area of accounting education, and the AAA has been at the heart of all of the major activities. The AAA's activities have multiplied many times over. However, it is interesting to question how much progress has actually been made toward the true objectives of accounting education. For example, the Accounting Education Change Commission (to be discussed in more detail in the next chapter) is the culmination of a number of committee reports and events. But, a comparison of the objectives of the AECC with the recommendations contained in the first (1970) report of the AAA Director of Education shows that little has been accomplished in the past quarter century. Zeff was calling for more dialogue on curricula, education philosophy, and teaching technique, for experimentation with new approaches to teaching, for improved accessibility to teaching resources, and for encouragement of research on the effectiveness of new approaches to teaching and of alternative curriculum patterns. Are these ideas just now becoming accepted? Was Zeff simply ahead of his time? Or have accounting educators simply been slow to change their traditional methodologies? It should be noted that in the Appendix to Zeff's first report, he described a committee similar to that of the AECC and a publication similar to *Issues in Accounting Education.*

One final point that should be noted is the fine job performed by Linda Sydenstricker, the AAA Meetings Coordinator. Linda is in charge of all local arrangements at all AAA conferences and seminars. As the number of seminars has increased, so has Linda's work, but she has performed beautifully despite the increased work load.

References

"American Accounting Association Annual Report," *The Accounting Review,* January, 1971, pp. 180-183.

American Accounting Association, "Report of the Doctoral Consortium and Fellowship Fund Evaluation Committee, April 6, 1981.

"Arthur H. Carter Scholarship Winners," *Accounting Education News,* October, 1989, p. 21.

"Award for Innovation in Accounting Education," *Accounting Education News,* November, 1990, p. 17.

Bostwick, Charles L., Dennis E. Grawoig, Edward G. Rodgers, and William J. Thompson, *Lecture and Problem Materials in Quantitative Methods* (Evanston, IL: American Accounting Association, 1971).

Bedford, Norton M., Letter to G. Kenneth Nelson, December 21, 1970.

Benjamin, James, "1986 Trueblood Seminars," *Accounting Education News*, June, 1986, pp. 4-5.

Committee to Prepare a Revised Accounting Teachers' Guide, *A Guide to Accounting Instruction: Concepts and Practices* (Cincinnati: South-Western Publishing Co., 1968).

"DH&S Foundation Contribution," *Accounting Education News*, November, 1989, p. 5.

"Doctoral Consortium," *The Accounting Review*, July, 1971, p. 608.

"Doctoral Programs Conference," *Accounting Education News*, June, 1983, pp. 1 and 4.

Edwards, James Don, "Comments from the President," *The Accounting Review*, January, 1971, pp. 165-166.

Horngren, Charles T., Interview by Dale L. Flesher, July 28, 1990.

Ijiri, Yuji, Letter to Robert L. Steele, January 17, 1983.

"Inaugural Senior Faculty Consortium," *Accounting Education News*, November, 1990, pp. 19ff.

Madden, Donald L., "Third Annual New Faculty Consortium," *Accounting Education News*, May, 1985, pp. 6-7.

"New Faculty Consortium," *Accounting Education News*, November, 1982, p. 11.

"1990 Seminar on Professionalism & Ethics," *Accounting Education News*, October, 1990, pp. 15-17.

Outstanding Educators: Fourteen Profiles, Sarasota: American Accounting Association, 1983.

"Outstanding Educators Movie," *Accounting Eduation News*, November, 1982, p. 13.

"The Trueblood Seminar—A Rewarding Experience," *Accounting Education News*, May, 1983, pp. 6-9.

"Trueblood Accounting Seminars Established," *Accounting Education News*, September, 1974, p. 1.

Zeff, Stephen A., Interview by Dale L. Flesher, December, 1989.

Zeff, Stephen A., "Report of the Director of Education," American Accounting Association, August, 1970.

Zeff, Stephen A., *The American Accounting Association: Its First 50 Years* (Englewood Cliffs, NJ: Prentice-Hall, Inc., 1966).

CHAPTER 7
EDUCATIONAL RESEARCH AND REFORMS

The AAA has been particularly active over the past quarter century in the area of educational research and reform. Most special publications were covered in the preceding chapter. This chapter covers the periodical publications in the area of educational research and a discussion of the Association's role in some of the controversial educational subjects of recent years, including accreditation, academic independence, minority issues, and educational reforms.

Education Section of *The Accounting Review*

For most of the past quarter century, *The Accounting Review* has had a special section for articles dealing with education research. In 1966 there was a "Teacher's Clinic" section which published the results of educational research, plus articles that could be viewed as sharing of successful teaching methods, and even an occasional "think piece." The "Teacher's Clinic" was replaced in January 1971 by a section entitled "Education Research." John A. Tracy (University of Colorado) was the first editor of the "Education Research" section. The January 1971 issue included an explanation that the new title reflected an evolutionary change of policy in that short notes were to be published separately from empirical studies. The "Education Research" column was subsequently eliminated with the April 1986 issue when the AAA decided to publish *Issues in Accounting Education* on a regular basis.

The editors of the "Teacher's Clinic" and "Education Research" sections during the past 25 years were:

1965-68	Neil C. Churchill (Harvard University)
1969-70	George C. Mead (Michigan State University)
1971-72	John A. Tracy (University of Colorado)
1972-74	Robert H. Strawser (Texas A&M University)
1975	Don T. DeCoster (University of Washington)
1976-78	Jay M. Smith (Brigham Young University)
1978-80	Edwin H. Caplan (University of New Mexico)
1980-84	L. S. Rosen (York University)
1984-86	Jack Kiger (University of Tennessee)
1986 (one issue)	Frank H. Selto (University of Colorado)

Despite the outlet provided by *The Accounting Review*, there were few articles published on the subject of accounting education, partially because of the low page count devoted to educational research. In fact, Harold Langenderfer had recommended as early as 1973 that a separate education journal was needed. As a result, the 1980-81 Committee on Accounting Education recommended that more publication outlets were needed.

Issues in Accounting Education

Research director James Fremgen discussed the recommendation of the Committee on Accounting Education at the March 1981 Executive Committee meeting. Four alternative possibilities were identified. These alternatives were to (1) restore the "Teacher's Clinic" column of *The Accounting Review*, (2) publish a fifth annual edition of *The Accounting Review* which would be devoted to educational articles, (3) start a separate journal, or (4) publish education-related articles in *Accounting Education News*. The latter publication had been established as the AAA newsletter in 1973. It was the consensus of the Executive Committee that Fremgen should pursue the possibility of a separate publication as a separate issue of *Accounting Education News*. An announcement calling for manuscripts was placed in *Accounting Education News* [Minutes, March 1981, p. 6]. By the August 1981 meeting, Fremgen had received seven papers for the experimental issue. Fremgen was replaced by Tom Burns in August 1981 and the special issue was scheduled for spring 1983 after Richard Murdock of Ohio State University had been appointed editor. At the March 1983 meeting, the Executive Committee decided to title the publication *Issues in Accounting Education—1983*. From the start, the journal was a quality publication; the acceptance rate the first year was 16 percent, and that declined to 15 percent the second year. Murdock remained as editor through the first three annual volumes.

Following approval by the Executive Committee at its March 1985 meeting, *Issues* became a semiannual publication in 1986 under the editorship of Robert W. Ingram (University of Alabama). *Issues* might never have become a regular, semiannual publication had it not been for the input from Council. A committee report recommending a new journal, which was ultimately to become *Accounting Horizons*, was sent to Council for its recommendation. Not only was Council enthusiastic about *Horizons*, but also encouraged the semiannual publication of *Issues*. Some past presidents feel that this insistence on the need for *Issues* was one of the most important contributions ever made by the Council.

Robert Ingram remained editor through 1988 and was replaced by Daniel L. Jensen (Ohio State University). *Issues* currently publishes three types of scholarly work: articles, instructional resources, and editorials and commentaries. In addition, the journal also publishes reviews of textbooks. The acceptance rate is now less than 10 percent. Fred Neumann (University of Illinois) was selected as the successor to Dan Jensen.

Accounting Accreditation

Accreditation has been an issue of concern throughout the past quarter century. Initially, the question was with regard to whether junior colleges should be accredited by AACSB. In 1969, director of education Stephen Zeff contacted William Flewellen regarding cooperation between the AAA and AACSB in developing criteria for future accreditation of accounting programs in junior colleges. However, the AACSB deans were not interested [Minutes, December, 1969, p. 4].

The Association's accreditation studies increased during the mid and late 1970s, much of it centering around the work of the Administrators of Accounting Programs Group (AAPG) (see chapter 5). At that time, the AICPA was proposing to establish an accreditation agency to administer the accreditation of accounting programs. This was brought about because the AACSB was opposing the separate accreditation of accounting programs and was objecting to any accreditation when the accounting function was not a part of the school of business. This issue produced one of the most controversial and unpleasant half hours in the history of Executive Committee meetings. At the August 1976 meeting, a motion was made by Floyd Windal and Fred Skousen that the AAA join with the AICPA in forming a new accreditation group. President Wilton Anderson was an ardent supporter of the idea. However, Thomas Dyckman, Charles Horngren, Robert Anthony, and Alfred Rapport, all from private MBA-oriented schools, were not enthusiastic about accounting accreditation, and certainly did not want it done by the AICPA. Dyckman and Anderson got into a loud argument over the issue. Finally, when one of the individuals opposed to the motion was out of the room, Anderson called for the vote. The vote ended in a 4-4 tie, but Anderson's vote resulted in the motion being passed. The next day, Dyckman tried to reopen the issue, but a furious and emotional Anderson was able to convince enough members that Dyckman was staging a personal attack, and the result was that the issue was not reopened [Dyckman, 1989]. The AAA participated with the AICPA in a joint committee on accreditation, but the new agency never came about.

The culmination of AAA's activities was a June 1978 Executive Committee meeting devoted solely to the discussion of accreditation. The committee concluded with one motion being approved: "The American Accounting Association shall cooperate in the AACSB Accreditation plan... by nominating AAA representatives to serve on the Accounting Accreditation Planning Committee" [Minutes, June, 1978, p. 2]. The AAPG opposed the AACSB proposal.

Another special Executive Committee meeting was held in December 1979. The major conclusion of that meeting was that the Executive Committee did not fully agree with the report of the 1978-79 AAA Accreditation Committee [Minutes, December 1979, p. 2].

Despite all of the activity with respect to accreditation, a 1988 committee report concluded that the AAA had never been effective in fully participating in the decision-making process because of the policy of not taking official positions. "While the AICPA and AACSB were very proactive, the AAA's role was nominal, despite the obvious importance of accreditation issues to the accounting academic community" ["Report By...," 1988, p. 13]. That criticism is no longer valid; in September 1990, President Al Arens held a conference call meeting of the Executive Committee to get approval to submit an accreditation committee report to the AACSB as the official position of the organization [Minutes, September, 1990].

Schism Committee

Throughout the past quarter century there has been a developing schism between academic accountants and practicing accountants. Much of that schism has been traced to the change in required accounting faculty credentials which occurred in 1967 and the subsequent differences between practitioner and educator academic backgrounds. When the AACSB required doctorates for accounting professors, it changed the nature of accounting education and resulted in conflicts between the academic and practitioner communities. These disagreements involve accounting accreditation, program culture and content, and research agenda and models [Bricker and Previts, 1990, p. 14].

To study the schism issue, the AAA established a schism committee under the chairmanship of John K. Shank (Harvard University) in 1979, the report of which concluded that very few research articles had an orientation toward accounting practice. At the same time, the emphasis on research was being increased, while the emphasis on quality teaching was declining. A 1987-88 professorial environment committee concluded that part of the reason for the schism was because academic research had deteriorated into exercises in mathematics without any emphasis on internal logic. The use of mathematical notation in articles made it difficult for practitioners to understanding articles because they had not been exposed to the same educational background as educators [Bricker and Previts, 1990, p. 12].

Nearly a decade later, a Committee on the Relationship Between Practitioner and Academic Communities, chaired by Wanda Wallace, concluded that a major obstacle in bringing the practice and academic communities together was the AAA policy of not taking official positions as an organization ["Report by...," 1988].

The Question of Academic Independence

The January 1979 issue of *Accounting Education News* contained an open letter to members from then president Maurice Moonitz on the subject of academic independence. Moonitz noted that he had been approached by an attorney looking for an expert witness to testify against a large CPA firm. The attorney had complained that it seemed impossible to find a willing academic because they were all fearful that the CPA firm would cut off funding of the school at which the academic taught. Moonitz questioned whether accounting academics had indeed sacrificed their independence in exchange for the financial support provided to accounting programs. He asked for responses from members as to whether a problem existed [Moonitz, January, 1979, p. 2].

Over 50 responses ensued, and Moonitz prepared a report on the matter which was discussed at the August 1979 Executive Committee meeting and published in the October 1979 *Accounting Education News*. The essence of the findings was that there was an apparent independence problem and the immediate result was the creation of an ad hoc committee to investigate the problem. That committee, chaired by Stephen Loeb (University of Maryland), was jointly appointed by 1979-80 president Don Skadden and 1980-81 president Joe Silvoso, because the committee was expected to work for two years before issuing its report. Other members of the committee were Robert K. Mautz (University of Michigan), Charles T. Horngren (Stanford), Robert W. Williamson (Notre Dame), William G. Shenkir (University of Virginia), and Jack L. Krogstad (Kansas State University). The committee's charge, along with a list of questions that the committee would be addressing, was published in the May 1980 *Accounting Education News* with the intent of familiarizing the membership with the issues prior to an open discussion of the issues at a concurrent session during the 1980 annual meeting in Boston ["Academic...," 1980, p. 3].

A summary of the committee's report was published in the January 1982 *Accounting Education News*. Without finding fault or making accusations, the committee concluded that academics should enter conflict situations with intellectual honesty and objectivity and not be influenced by any real or supposed obligation to another party. Also, academics should not shun a situation merely because of appearances of lack of independence. It was acknowledged that the potential existed for academic independence to be questioned when financial considerations were involved, but the committee concluded that no code of ethics for academic accountants was necessary since professors are subject to the same codes as practicing accountants ["Summary of...," 1982, pp. 3-7].

A similar issue arose in 1987 when there were rumors that certain textbook publishers had been paying commissions or kickbacks to professors and departments to induce adoption of specific textbooks. At the November 1987 Executive Committee meeting, a letter was approved to be sent to all known accounting textbook publishers regarding the purported practices. Later, the American Assembly of Collegiate Schools of Business passed a resolution in support of the AAA on the matter ["Accounting...," 1989, p. 3].

Bedford Committee

In 1984 a committee was appointed by President Doyle Williams with the impressive title of Committee on the Future Structure, Content, and Scope of Accounting Education. Because its chairman was Norton Bedford (University of Illinois), the committee has come to be known as the Bedford Committee. The committee had an imposing list of members including three past AAA presidents, at least one future president, the comptroller general of the United States, and the president of the National Association of Accountants. The committee studied the features of the expanding accounting profession and the current state of accounting education. A report was issued in 1986. The major conclusion was that accounting education would require major reorganization before the turn of the 21st century. The committee's recommendations were intended to serve as broad guidelines and to provide direction for those who wish to initiate changes in accounting education. There were 28 recommendations including ten on the future scope, content, and structure of accounting which addressed the most comprehensive concerns of the committee. Other recommendations were associated with the teaching process, faculty responsibilities, administration, accreditation, professional examinations, and economics of accounting education ["Future...," 1986].

Subsequently, presidents Ray Sommerfeld and William Beaver appointed committees to follow up on the Bedford Committee report and to examine the accounting education environment and curriculum. The reports of these follow-up committees were published in 1989 as Accounting Education Series Monograph No. 10, edited by Joseph J. Schultz, Jr. [Schultz, 1989].

Big-8 Initiative and the AECC

Following the Bedford Committee report, the next major issuance on the subject of accounting education was what has been called the "Big-8 White Paper," but formally titled *Perspectives on Education: Capabilities for Success in the Accounting Profession*. That publication, which noted a concern about the quality and number of accounting graduates, basically agreed with the conclusions of the Bedford Committee. As a part of the publication, the Big-8 CPA firms pledged up to $4 million to support the design and implementation of innovative curricula, new teaching methods, and supporting materials that will equip graduates for success. The AAA was challenged to provide leadership in bringing about that implementation.

This leadership was evidenced by the establishment of the Accounting Education Change Commission (AECC) at the April 1989 Executive Committee meeting. The objective of the AECC is to foster changes in the academic preparation of accountants. The AECC operates under AAA oversight, but has significant autonomy with respect to the conduct of its activities. Doyle Williams, a past AAA president, was named as chairman of AECC, and Gary Sundem (University of Washington), a future president, was named as executive director. At its first meeting, the AECC agreed to accept the basic thrust of the two aforementioned committee reports. It is too early to evaluate the overall contributions of the AECC, but following a quarter century of study, discussion, and debate, the AECC represents a major attempt at action with respect to changes in the accounting curriculum.

CPA Examination

During 1985 and 1986, the Executive Committee adopted a pair of motions concerning the CPA examination. Basically, one motion endorsed a portion of a report by the 1983-85 AAA Committee on Professional Examinations which was critical of trends in the scope and content of the CPA exam, namely that the coverage was too narrow and the manner of questioning candidates too mechanical. The other motion addressed the timing of the exam in relation to the academic year. The latter motion, a personal favorite of then president Stephen Zeff, encouraged the AICPA to change the date and frequency of the CPA examination. Zeff was a supporter of a once-a-year exam to be given in August. As an interim measure, he urged states to adopt the Louisiana formula of not permitting candidates to apply for the exam until they had completed their educational requirements. This would preclude seniors from taking the exam during their final semester of college [Zeff, 1986, pp. 1-3].

In 1988, the AAA Executive Committee spent considerable time debating the AICPA proposal to change the CPA examination to an all objective format. The result was formal support of the need for using essay questions. Of more importance than the conclusion of this debate was the fact that the Executive Committee took an official position on the subject of essay questions on the CPA examination [Beaver, 1988, p. 5]. Previously, the Executive Committee had been reluctant to take official positions on behalf of the membership.

Minority Considerations

A person looking at pictures of past AAA presidents would immediately note that they are all white males. However, in the 1960s and early 1970s, there were few blacks in the teaching profession and few women. As late as 1970, there were only three black women with doctorates in accounting. Despite the few blacks and women in the profession, the AAA has long worked to increase the numbers of these groups. There has even been some minority representation on the Executive Committee. Two vice presidents have been black. The first was William Campfield. Later, Sybil Mobley became the first black woman on the Executive Committee in 1979.

Throughout this quarter century the AAA has worked to recruit blacks into the profession. There have been committees with this as their charge, and there has been cooperation with the AICPA on fellowship programs for minorities who want to teach accounting. In 1974, the AICPA and AAA jointly submitted a proposal to the U.S. Department of Health, Education and Welfare (HEW) for a fellowship program. A grant resulted and during 1975-76 grants were made to six black doctoral students. The students received a $7,500 grant, and their predominantly black employers were each granted $16,000 as a salary supplement for visiting scholars to replace the doctoral candidates. The $16,000 came from HEW as did $4,000 of the grant to each individual. The AICPA gave each candidate $3,000. The costs to AAA included $1,000 paid to the AICPA for administration of the program, and $500 given to each candidate. Two individuals joined the program in 1976, and eight more in 1977 [Gerhardt, 1977].

Summary

The AAA has been at the forefront of changes in accounting education during the past quarter century. Admittedly, the Association has not been alone. There have been many contacts with the AICPA, NAA, AACSB, and other organizations. In fact, for several years there were annual meetings wherein the AAA president met with the presidents of the other major accounting organizations; these were known as "Summit" meetings and were eliminated in the late 1970s because there was a question as to whether there were any benefits to be gained. There was also during the late 1960s an annual "pinnacle" meeting in which the AAA president met with the AICPA president. The Careers Council of the 1960s was another joint effort with other accounting organizations to publish materials to recruit students into the accounting profession. The large CPA firms and many industrial firms also must be lauded for the financial contributions they have made to many AAA educational programs. Without these contributions, the AAA would not have been nearly so active over the past 25 years. These outside contributions that today total hundreds of thousands of dollars a year are a product of this quarter century; in 1965, total firm donations to the AAA came to about $1,000. Another organization with which the AAA has worked closely has been Beta Alpha Psi; that organization now rents an office in the AAA building in Sarasota.

In summary, the AAA has done much in the way of accounting education in the past 25 years—both in terms of direct activity and as a coordinator of such activities. There are other accounting organizations which are concerned with accounting education, but the AAA is the primary education association concerned with accounting.

References

"Academic Independence," *Accounting Education News*, May, 1980, p. 3.

"Accounting Textbook Inducements," *Accounting Education News*, October, 1989, p. 3.

Beaver, William H., "CPA Examination," *Accounting Education News*, October, 1988, p. 5.

Dyckman, Thomas R., Interview by Terry K. Sheldahl, December 9, 1989.

"Future Accounting Education: Preparing for the Expanding Profession," *Issues in Accounting Education*, Spring, 1986, pp. 168-195.

Gerhardt, Paul L., Letter to Executive Committee, July 29, 1977.

Minutes of the American Accounting Association Executive Committee, 1965-1990.

Moonitz, Maurice, "An Open Letter to All on a Matter of Independence," *Accounting Education News*, January, 1979, p. 2.

Moonitz, Maurice, "A Report on a Matter of Independence," *Accounting Education News*, October, 1979, pp. 3-7.

"Report By The Committee On The Relationship Between Practitioner and Academic Communities," American Accounting Association, April, 1988.

Schultz, Joseph J., Jr., *Reorienting Accounting Education: Reports on the Environment, Professoriate, and Curriculum of Accounting* (Sarasota: American Accounting Association, 1989).

"Summary of the Report of the Committee on Academic Independence," *Accounting Education News*, January, 1982, pp. 3-7.

Zeff, Stephen A., "President's Message," *Accounting Education News*, June, 1986, pp. 1-3.

CHAPTER 8
RESEARCH ACTIVITIES

The research activities of the Association during the past quarter century have been many and varied. These activities are examined by looking at the contributions of the research directors, the activities of the journal editors, and the winners of the Association's major research awards. Charles Horngren was the last director of research of the second quarter century and had inaugurated a new thrust in the research arena. In 1966, Horngren turned the position over to Robert K. Jaedicke of Stanford University. The Association's major publication of 1966 was *A Statement of Basic Accounting Theory* (ASOBAT), which all members had received, and for which an extra 9,000 copies were printed to be sold at $2 per copy. Arthur Andersen & Co. had provided a grant of $1,000 toward the total cost of $19,577. This latter figure was broken down into $9,836 for committee expense, largely travel, and $9,741 for printing and postage [Minutes, August 20-21, 1966, p. 8]. Charles T. Zlatkovich chaired the ASOBAT Committee. Stephen Zeff, 1985-86 president, later called ASOBAT "one of the most important studies ever issued by the Association, both in terms of impact on the literature and on standard-setting (Zeff, 1991). Although ASOBAT was a monumental accomplishment, it was one of the last major research projects that could be classified as committee research. Thus, 1966 represented a turning point in Association research as there was a move toward individual research away from committee research.[1] Of course, cost could have been a partial reason for that move. The cost of $9,836 for committee expense on ASOBAT can be compared to individual research grants which were typically much less. Consequently, the work of 1966-68 research director Robert Jaedicke differed somewhat from that of his predecessors.

One other example of committee research was published in 1977; that was *Statement on Accounting Theory and Theory Acceptance*, which was initially commissioned by the Executive Committee in 1973. The charge was to prepare a statement that would provide the same type of survey of current thinking on accounting theory as ASOBAT had done in 1966. The Committee on Concepts and Standards for External Financial Reports, chaired initially by Kermit Larson and later by Lawrence Revsine, found that the basic disciplines used by accounting theory had been altered considerably in only a decade. As a result, the committee could not come up with easy theoretical answers to the problems facing accounting. For that reason, the 1977 report was not a statement of accounting theory, but a statement about accounting theory and theory acceptance [Statement on..., 1977, p. ix]. One conclu-

---

[1]This is not to say that committee research was eliminated. The Association continued to publish committee reports as supplements to *The Accounting Review* throughout the decade of the 1970s. Many of these reports dealt with projects that had research elements.

sion reached by the committee, which probably had already been recognized by many AAA leaders, was that appointed committees cannot efficiently conduct research [p. 49]. Nils Hakansson, in a review article that was critical of the committee report, laid to rest the subject of committee research with the following lines:

> there are many reasons why I am far from convinced that the committee approach makes sense even in summarizing the state-of-the-art. There are few areas in which decentralized production decisions generated by self-interest can be surpassed, and scholarly activities is most likely not one of them. And when most of the members of a committee keep citing their own minor works, the effect is somehow more pronounced than when single or multiple authors do so [Hakansson, 1978, p. 724].

The Research Directors

Robert K. Jaedicke, 1966-68. Jaedicke's (Stanford University) first year was rather slow because, at that time, projects initiated by earlier research directors were still under the domain of those individuals. Thus, Jaedicke's concerns were primarily with the lack of progress being made by researchers who had been commissioned by Charles Horngren. The thrust of Jaedicke's first annual report was that AAA should work closely with NAA, FEI, and AICPA to encourage those organizations to use AAA members in their research projects [Minutes, August 26-27, 1967, p. 5]. During 1967, Jaedicke began advocating summer research support to encourage research. In March 1968, the Executive Committee approved the granting of summer research grants on the basis of research proposals submitted to the director of research. These grants could not exceed $5,000. Neither could the total of all grants exceed $20,000 per year [Minutes, March, 1968].

David Solomons, 1968-1970. Solomons (University of Pennsylvania), in 1969, oversaw the first publication in the research monograph series which had been initiated in 1965 by then research director Horngren. The series was slow to gain fruition, partly because researchers were not able to devote enough time to such projects due to a shortage of funds [Minutes, December, 1968]. The first publication in the series was *Investment Analysis and General Price-Level Adjustments: A Behavioral Study* by future president Tom Dyckman of Cornell University. Charles Horngren had commissioned Dyckman's study. Another 1969 publication was Baruch Lev's (University of Chicago) study entitled *Accounting and Information Theory*, which had been commissioned by Solomons. Also published in 1969 was *The Allocation Problem in Financial Accounting Theory*, by Arthur L. Thomas (McMaster University). Jaedicke had commissioned the Thomas study. Thomas' monograph was one of the most favorably reviewed and best selling of any in the entire monograph series.Thomas concluded that it was impossible to defend any method of allocation and that either allocation theory had to be improved or a new approach to financial accounting had to be adopted that would eliminate the need for allocations.

Hector Anton, 1970-1972.   Anton (Berkeley) was the first research director to have official responsibility over projects commissioned by his predecessors. In 1970, Studies in Accounting Research (SAR) No. 4, by Horwitz, was published; SAR No. 5 followed in 1972—both of these commissioned by Anton's predecessors.

Robert R. Sterling, 1972-1974.   Sterling (Rice University) commissioned seven research projects during his first year. He also published several monographs which had been commissioned by his predecessors. Sterling was very enthusiastic about the role of the research director—so much so that he turned down the editorship of *The Accounting Review*. Sterling's major disappointment was the lack of appreciation of research outside one's own specialization. For instance, he urged Maurice Moonitz to publish his recollections of the history of standard setting, but found that accountants who professed some knowledge of history rejected the project because it did not meet their peculiar definition of history. Sterling tried to explain that Moonitz had been a participant on standard setting and if they wanted to call this memoirs instead of history, it did not matter, but the project was worth publishing. He found other examples of the same sort of thing: a person interested in researching "x" would argue against the publication of the work of someone who was researching "y." Sterling feels that it was the result of this narrow view that the number of monographs declined after he left the research directorship. Sterling has stated that in most cases, the reason given for rejection was "lack of quality," when in fact it was the lack of breadth of the people who were judging the monograph. "I fear that the same sort of thing continues today. Much of what we see published in the journals is the result of what is a rather narrow definition of what is acceptable research rather than a broad view of research for improvement of the state of knowledge" [Sterling, 1991]. Sterling also had the unique situation of having the recipient of a 1969 research grant offer to repay that grant to the AAA because the findings were not publishable. The Executive Committee voted to decline the offer with the argument that "funds for research are somewhat of a gamble and that one never knows how true research is going to work out" [Minutes, August 13, 1973, p. 8].

K. Fred Skousen, 1974-1976.   Skousen (Brigham Young University) published monographs by Maurice Moonitz, Arthur Thomas, and Yuji Ijiri during his first year. The Thomas study, entitled *The Allocation Problem: Part Two*, was a sequel to SAR No. 3. Skousen also commissioned several studies. Unlike many of his predecessors, Skousen committed his entire budget for support of research subjects. Others had returned money from their budget. Skousen remembers that the major difficulties in the job were obtaining research proposals from prospective researchers, and getting the recipients of grants to finish their projects on a timely basis [Skousen, 1991].

Thomas R. Dyckman, 1976-1978.   Dyckman (Cornell University) commissioned three studies out of the 17 proposals received during his first year in office. Subsequently, these commissions resulted in SAR Nos. 17 and 30. Dyckman published SAR No. 14 by Robert Jensen and SAR No. 15 by Hiroyuki Itami, both commissioned by his predecessors. The Jensen monograph carried the intriguing title of *Phantasmagoric Accounting.* Dyckman lamented the fact that many researchers had not completed their studies even after five years. In 1977, he gave all researchers with studies more than one year past their expected completion date, one more year to submit a manuscript; otherwise their project would be cancelled [Minutes, August, 1977, p. 3]. Among those cancelled was one authored by a future AAA president. Dyckman felt that the previous methods used to commission studies had not been effective since few had ever been completed. Thus, he looked for people who were doing recognized research and encouraged them to write monographs. A number of people agreed, but nothing ever came of those commitments. Dyckman later observed that many researchers do not want to get involved with a monograph because they take longer than articles to complete, and may not receive any more credit (or even as much credit) as an article [Dyckman, 1989].

Eugene Comiskey, 1978-1980.   Comiskey (Purdue University) was particularly concerned about the low number of proposals received. His response was to solicit proposals through letters to professors throughout the nation. As had been true of at least one of his predecessors, Comiskey urged the Executive Committee to extend the term of office of the research director to three years; the committee did not agree [Minutes, March, 1980, p. 4]. Comiskey commissioned five projects during his tenure, three of which were subsequently published.

A. Rashad Abdel-khalik, 1980-82.   Abdel-khalik (University of Florida) had several innovative ideas which did not reach fruition due to concerns from the Executive Committee. He did operate under a system which required researchers to refund their grant stipends if no results were achieved. Abdel-khalik commissioned six projects during his tenure—all of which were subsequently published. Obviously the requirement that grant funds be returned was a powerful incentive for researchers to complete their projects.

Theodore J. Mock, 1982-1984.   Mock (University of Southern California) commissioned four studies near the end of his first year, and elected to publish Carl T. Devine's *Essays in Accounting Theory* as a five-volume SAR No. 22. President Yuji Ijiri was responsible for getting Devine to allow his work to be published in the series; apparently some editor of *The Accounting Review* had miffed Devine in the past, and as a result he was reluctant to have his materials published. Following a visit from Ijiri, Devine agreed to let AAA publish his work. The 59 essays, originally published by the author in mimeograph form, combine to form an encyclopedia of accounting foundations. Mock also compiled a list of research funding opportunities, from sources outside of AAA, which was published in *Accounting Education News*, and was of major benefit to seekers of grants. In retrospect, Mock has verbalized an aspect that others have hinted at—namely that the director of research has little responsibility or opportunity to do much because of a limited budget (normally no more than $20,000 per year) [Mock, 1991].

William R. Kinney, 1984-1986. Kinney (University of Texas) kept his costs low the first year by accepting only one research proposal out of the twelve received, and that one required no funding. However, he did publish four SARs (21 through 24) in 1985, all of which had been accepted by previous research directors. The one monograph which Kinney authorized was No. 26 by Cushing and Loebbecke, which was entitled *Comparison of Audit Methodologies of Large Accounting Firms.* The monograph had been commissioned by the AICPA, but one of the Big-8 firms objected to some of the findings and the AICPA would not publish the monograph. Kinney and his review committee liked the manuscript and committed the AAA to publish the material. It turned out to be a good seller among both practitioners and professors. Thus, the AAA got the benefit of the AICPA's financial support and the community got to read what had become a controversial document [Kinney, 1991].

Barry E. Cushing, 1986-1988. Cushing (Penn State University) updated the work on the list of research funding opportunities started by his two predecessors, and published the available grants in *Accounting Education News.* He also edited and published a volume entitled *Accounting and Culture,* consisting of the plenary session papers from the 1986 annual meeting.

Shyam Sunder, 1988-1990. Sunder (Carnegie Mellon University) supported a policy prohibiting the payment of research stipends (paying for members' time to conduct research) without special permission from the Executive Committee and to limit reimbursable expenses to $5,000. Reimbursements will not be made until after a manuscript has been submitted [Minutes, April, 1990, p. 4]. Thus, the quarter century began with the AAA inaugurating a new program of research funding to encourage professors to conduct research during the summer months, and ended with the elimination of that program. Part of the reason for the elimination of research stipends was the fact that large CPA firms and their foundations were providing research grants far larger than could be funded by the Association. At this same time the Executive Committee addressed the subject of paying honoraria and stipends to members for other services to the Association. It was decided that the AAA should not pay any member for his or her time, including any time spent on conducting research [Sunder, 1991].

Sunder also supported the policy of encouraging authors of AAA-published research to make their data available for use by others in extending or replicating results. Authors of articles which report data-dependent results should footnote the status of data availability and how the data may be obtained [Minutes, April, 1989, p. 3]. William W. Cooper, director of publications was the major force behind the latter initiative.

Nicholas Dopuch, 1990-1992. Dopuch (Washington University), a two-time winner of the AAA/AICPA Accounting Literature Award and a recipient of the AAA Outstanding Educator Award, has been perhaps the most decorated of research directors. It is still too early in his term to analyze his contributions to the AAA's research efforts.

Research Journals
The Accounting Review

In 1966, *The Accounting Review* was being edited by Wendell Trumbull (Lehigh University). It was not a peer reviewed journal; Trumbull made all editorial decisions without benefit of outside reviewers. For this effort, he received a $3,300 honorarium from the Association. An editorial review board did begin functioning in 1966, marking another new era in the operations of the Association.

The beginning of AAA's third quarter century was in the midst of a transition period for *The Accounting Review*. Shortly after it was founded in 1926, the journal was predominantly a theoretical journal. A decade later it became a vehicle for promoting changes in accounting practice and theory. Finally, in the late 1950s and throughout the 1960s it became an outlet for empirical studies in which theory based on accounting principles became less important [Chatfield, 1975, p. 1]. During the 1960s the journal began publishing more articles by nonaccountants on topics such as how methods and ideas from other disciplines could be used to solve accounting problems. These topics included mathematical model building, organization theory, linear programming and Bayesian analysis.

Charles Griffin (University of Texas) became editor in 1967 and served three years. The acceptance rate for 1967 was about 40%. Griffin received a $3,300 honorarium the first year, and $3,600 in his second year, which he transferred to his department at the University of Texas to cover the cost of an additional secretary [Zlatkovich, 1990]. In 1970, when the Executive Committee learned that Griffin was using his honorarium for secretarial help, it voted to give him an extra $3,000; however, Griffin declined the money saying that it was a privilege to serve as editor. He did, however, recommend that future editors should receive expense reimbursements in addition to their honoraria [Minutes, April, 1970, p. 5]. The acceptance rate declined in 1969 to 24% as only 56 papers were accepted from the 251 received.

Eldon Hendriksen (Washington State University) served two years as editor (1970-72). The question of readability was brought up by Charles Zlatkovich at the December 1970 Executive Committee meeting when it was noted that since the bulk of the membership is not from academia or from the larger universities, the material in the journal should be kept at a more basic level. Hendriksen agreed, but stressed the need to keep the general quality high. In an attempt to address the criticisms of the journal, a study was conducted by Edwin Caplan and Charles Griffin to assess the membership view of *The Accounting Review*. All members of the AAA were sent a 10-question questionnaire. After examining over 2,800 responses, the authors concluded that:

> the present character and content of *The Accounting Review* is considered acceptable and appropriate by the majority of its readers. It is likely that, given the extreme diversity of interest of the readership, the journal is doing about as well as can be expected in this regard. Despite the criticisms raised by a number of Association members, there does not appear to be any widespread serious dissatisfaction with the publication [Caplan, 1972, p. 15].

During Hendriksen's first year he received 365 manuscripts; this did not include the manuscripts received by John Tracy, the editor of the "Education Research" section [Minutes, August, 1971]. Hendriksen's honorarium for 1971-72 was $3,900, plus $3,000 for expenses. Editorial activities consumed 40 to 50 hours per week of Hendriksen's time [Minutes, August 22, 1971, pp. 3-4]. To help alleviate the work load, the Executive Committee at its December 1971 meeting approved an additional budget of $5,000 per year for editorial assistance.

Thomas F. Keller (Duke University) was editor from 1972-75. Keller took over at a time when there was considerable member dissatisfaction with the readability of the journal. President Sprouse assigned this problem to the Long-Range Planning Committee which subsequently performed a thorough readership survey. The acceptance rate during Keller's first year was 20% of the 400 manuscripts received. During his final year as editor, Keller made the decision to not publish any manuscript which was published in the Proceedings of the annual meeting (which were published for the first time in 1975).

Don DeCoster (University of Washington) served two years as editor (1975-77). He processed nearly 800 manuscripts during this two-year period. One of DeCoster's claims to fame was that he held two Ph.D.s, which probably gave him greater insight with regard to manuscripts that applied ideas from other disciplines to accounting. Submissions increased considerably during DeCoster's term, while the acceptance rate declined to 13%. The "News Notes" section which had appeared at the back of the journal for half a century was, at the suggestion of DeCoster, moved to *Accounting Education News* in 1976.

During 1975-76, a committee chaired by Maurice Moonitz examined the editorial policy of the journal and considered the merits of additional publication outlets. The journal's areas of interest were defined to include auditing, financial accounting, governmental and not-for-profit accounting, managerial accounting, information systems, and tax accounting. The committee felt that the AAA should publish the results of research employing sophisticated methodologies, as well as those using older, more familiar modes of thought and reasoning. The committee recommended that a submission fee be initiated and that each article be preceded by an abstract ["Report of...," 1976, p. 4]. The recommendation regarding the need for an abstract was the result of a member survey conducted by committee member Rashad Abdel-khalik [1976, p. 616]. With respect to other publication outlets, the committee recommended the continuation of all AAA publications, and the addition of a separate issue of *The Accounting Review* for publication of selected papers presented at AAA annual meetings ["Report of, 1976...," p. 4]. The latter recommendation was never acted upon.

Stephen Zeff (Rice University), who served as editor from 1977-82, held the position the longest of anyone this quarter century. He saw many changes during his tenure including the installation of a $25 submission fee on July 1, 1977, and a by-laws change that permitted the editor to serve longer than three years, but without privilege of serving on the Executive Committee for more than three years. In addition, during Zeff's tenure (1979) the position of editor-elect was approved, an idea that had been recommended several years earlier. Zeff's first year as editor was the roughest in that he was out of the country during the first month of his term serving as the AAA Distinguished International Lecturer in Latin America. When he returned, he moved to Boston where he was a visiting professor at Harvard University. At the end of that year, he moved to Rice University (he had previously been at Tulane).

During his first year as editor, Zeff analyzed the characteristics of submitting authors and found that 60% of all manuscripts were authored by an assistant professor, with another 15% coauthored by at least one assistant professor. Zeff concluded that assistant professors were the primary contributors to accounting literature. Zeff introduced several innovations in the journal while serving as editor. In 1978, he began publishing tables of contents from other research journals on a reciprocal basis. This arrangement continued through 1990 when the service was switched from the journal to *Accounting Education News*. Zeff also added a "Notes" section in 1979 and a "Financial Reporting" section [Zeff, 1978, p. 7]. In 1980, an index was compiled and published of all articles that had appeared in *The Accounting Review* from the first issue in 1926 through 1978. The compilers were Gary John Previts and Bruce Committe of the University of Alabama.

Gary Sundem (University of Washington) served as editor for four years from 1982-86. Sundem also was the first to serve as editor-elect (1981-82). Sundem was fortunate in having a cadre of associate editors to assist him, especially since he suffered a severe illness during his first year in office which prohibited him from fully fulfilling his editorial duties. However, the journal continued to be published on time [Zeff, 1989].

William Kinney (University of Texas) took over as editor in 1986 and served for three years. During his first year he received a $14,000 honorarium, the same as had been paid to his predecessor. By 1989, the annual honorarium to the editor had reached $25,000. The Executive Committee viewed this with alarm, particularly because the AAA now had three journals, and to provide such honoraria to three editors would be prohibitive. Therefore, the honoraria were eliminated following the 1988-89 fiscal year. This does not mean that editors do not have an expense allowance; they do. The rise to $25,000 came about because Robert Mautz was offered $25,000 to get him to be the editor of *Accounting Horizons.* Mautz was retired and did not have available the resources of a university employer; thus, the money was considered by the Executive Committee to be of more importance to him. However, it was felt that it would be unfair to give Mautz $25,000 and not also give that level of honorarium to the editor of *The Accounting Review.* Kinney kept only $10,000 of his $25,000 honorarium; the remainder went to associate editors. In retrospect, Kinney feels that there is no need for such honoraria because the honor of being editor is sufficient reward. Kinney was often criticized while editor. Just as Stephen Zeff had broadened the contents of the journal, Kinney was accused of narrowing the focus. The Executive Committee refused to officially discuss these complaints with Kinney, but encouraged the director of publications to do so. Kinney later stated that he was never aware of any criticism about his editorship. Some of the areas he was accused of slighting ultimately turned out to have among the highest acceptance rates of any manuscripts submitted to the journal [Kinney, 1991]. During Kinney's term, the Executive Committee voted to eliminate the "Educational Research" and "Financial Reporting" sections from *The Accounting Review* since the new journals *Issues in Accounting Education* and *Accounting Horizons* published similar type articles. The official motion was "An Association journal should not devote a special section to subject matter that falls within the purview of another Association journal" [Minutes, April, 1987]. Kinney did add a new section entitled "Small Sample Studies," a section that was later eliminated by his successor.

Rashad Abdel-khalik (University of Florida) knew about the elimination of honoraria when he assumed the editorship in 1989. Abdel-khalik claims to be open to any type of accounting-related manuscript and has seemingly broadened the focus of the journal back to what it was before his predecessor, although only after considerable complaints from tax professors during the first few months of the new editor's term. For over a decade, *The Accounting Review* had published tables of contents from other research journals on a reciprocal basis when space was available. In 1990, the Executive Committee voted to move these service announcements to *Accounting Education News* [Minutes, August, 1990, p. 2].

Individuals with a variety of research outlooks have held the position of editor of *The Accounting Review* during the past quarter century, and perhaps that is as it should be. Historically, it was a very powerful position within the organization, although that has changed slightly since the editor no longer serves on the Executive Committee. Despite the allure of the position, it is hard work—most past editors have claimed that they worked about 40 hours per week on their editorship duties. Because of the work load, some individuals have had to decline the editorship because of other commitments. Despite the honor of the position, there have been at least three instances in the past 25 years where individuals have declined the editorship.

Accounting Horizons

A new journal, *Accounting Horizons,* was established in 1986 under the editorship of Robert Mautz. *Horizons* is intended to be a high quality research journal, but appeal to practitioners as well as academicians because it emphasizes applied research. Mautz served as editor during the journal's first two years. He was replaced following the 1988 issues by the editorial team of John C. Burton (Columbia University) and Robert Sack (University of Virginia), who served through 1990. Jerry Arnold (University of Southern California) was selected in 1990 to replace Burton and Sack. *Accounting Horizons* has received many laudatory comments over its short history, but it is uncertain whether it has achieved one of its goals—that of bringing practitioners back into the membership fold.

Awards
Notable Contributions To Accounting Literature

The AAA, in association with the AICPA, grants an annual award for a notable contribution to accounting literature. An AAA committee makes recommendations to the AICPA of articles and books which should be considered for the award. An AICPA committee then makes the final selection. At times, particularly in the late 1970s and early 1980s, this award has been a political issue between the two sponsoring organizations. On at least two occasions, the AICPA selection committee chose a winner that was not nominated by the AAA committee. The winners of the award, which was first given in 1966, are listed below.

1966 **R. J. Chambers**, *Accounting Evaluation and Economic Behavior.*

Y. Ijiri and R. K. Jaedicke, "Reliability and Objectivity of Accounting Measurements," *The Accounting Review.* July, 1966.

W. J. Vatter, "Accounting for Leases," *Journal of Accounting Research,* Autumn, 1966.

Thomas R. Dyckman, "On the Effects of Earnings—Trend, Size and Inventory Valuation Procedures in Evaluating a Business Firm," *Research in Accounting Measurement.* 1966.

1967 **Yuji Ijiri,** *The Foundation of Accounting Measurement,* Prentice-Hall, 1967.

Joel S. Demski, "An Accounting System Structured on a Linear Programming Model," *The Accounting Review,* October, 1967.

1968 **Robert R. Sterling,** "The Going Concern: An Examination"" *The Accounting Review,* July 1968.

William H. Beaver, "Market Prices, Financial Ratios, and the Prediction of Failure," *The Journal of Accounting Research,* Autumn 1968.

1969 **Jack Gray and John K. Simmons,** "An Investigation of the Effects of Differing Accounting Frameworks on the Prediction of Net Income," *The Accounting Review,* October 1969.

David Solomons, *Divisional Performance: Measurement and Control,* Financial Executives Research Foundation, Inc., 1965. (Later published by Richard D. Irwin, Inc.)

1970 **Robert K. Mautz,** "Financial Reporting by Diversified Companies" (New York: Financial Executives Research Foundation, 1968).

Joel S. Demski and Gerald Feltham, "The Use of Models in Information Evaluation," *The Accounting Review,* October 1970.

Arthur L. Thomas, "The Allocation Problem in Financial Accounting Theory — Studies in Accounting Research #3," American Accounting Association, 1969.

1971 **C. West Churchman,** "On the Facility, Felicity, and Morality of Measuring Social Change," *The Accounting Review,* January, 1971.

Yuji Ijiri and Robert S. Kaplan, "A Model for Integrating Sampling Objectives in Auditing," *Journal of Accounting Research,* Vol. 9, No. 1, Spring, 1971.

1972 **Richard Mattessich,** "Methodological Preconditions and Problems of A General Theory of Accounting," *The Accounting Review,* July, 1972, pp. 469-487.

1973 **Robert R. Sterling,** "Accounting Research, Education and Practice," *The Journal of Accountancy,* September 1973, pp. 44-52.

1975 **Nicholas Gonedes and Nicholas Dopuch,** "Capital Market Equilibrium, Information Production, and Selecting Accounting Techniques: Theoretical Framework and Review of Empirical Work," Supplement to Volume 12, *Studies on Financial Accounting Objectives: 1974, Journal of Accounting Research,* pp. 48-129.

1976 **Yuji Ijiri,** "Theory of Accounting Measurement," Studies in Accounting Research No. 10, American Accounting Association, Sarasota, Florida, 1975.

1977 **Robert G. May and Gary L. Sundem,** "Research for Accounting Policy: An Overview," *The Accounting Review,* October, 1976, pp. 747-763.

1978 **Peat, Marwick, Mitchell & Co.,** *Research Opportunities in Auditing.*

1979 **William H. Beaver,** "Current Trends in Corporate Disclosure."

Ross L. Watts and Jerold L. Zimmerman, "Towards a Positive Theory of the Determinants of Accounting Standards," *The Accounting Review,* January, 1978, pp. 112-134.

1980 **Eldon S. Hendriksen,** *Accounting Theory,* Richard D. Irwin, Inc., 1977.

Ross L. Watts and Jerold L. Zimmerman, "The Demand for and Supply of Accounting Theories: The Market for Excuses," *The Accounting Review*, April, 1979, pp. 273-305.

1981   **Leo Herbert,** *Auditing The Performance of Management*, Wadsworth, Inc. (Lifetime Learning Publications), Belmont, Calif., 1979, pp. xvii, 476.

George Foster, "Accounting Policy Decisions and Capital Market Research," *Journal of Accounting and Economics*, 1980, pp. 29-62.

1982   **Nicholas Dopuch and Shyam Sunder,** "FASB's Statements on Objectives and Elements of Financial Accounting: A Review," *The Accounting Review*, January, 1980, pp 1-21.

Michael W. Maher, "The Impact of Regulation on Controls: Firms' Response to the Foreign Corrupt Practices Act," *The Accounting Review*, October 1981, pp. 751-770.

1983   **William H. Beaver,** "Financial Reporting: An Accounting Revolution," Prentice-Hall, Inc. 1981

George J. Foster, "Financial Statement Analysis," Prentice-Hall, Inc. 1978

1984   **Richard Leftwich,** "Evidence of the Impact of Mandatory Changes in Accounting Principles on Corporate Loan Agreements," *Journal of Accounting and Economics*, March 1981

1985   **Robert Libby,** "Accounting and Human Information Processing: Theory and Applications," Prentice-Hall, Inc.,1981

William R. Kinney, Jr. and William L. Felix, Jr. "Research in the Auditor's Opinion Formulation Process: State of the Art," *The Accounting Review*, April 1982

1986   **Carl T. Devine,** "Essays In Accounting Theory," American Accounting Association, 1985

1987   **Robert S. Kaplan,** "Measuring Manufacturing Performance: A New Challenge For Managerial Accounting Research," *The Accounting Review*, October 1983

1988

1989   **Mark A. Wolfson,** "Empirical Evidence of Incentive Problems and Their Mitigation in Oil and Gas Tax Shelter Programs," in *Principals and Agents: The Structure of Business*, Harvard Business School Press, 1985

1990   **Paul M. Healy,** "The Effects of Bonus Plans on Accounting Decisions," *Journal of Accounting and Economics*, 1985.

Wildman Medal Award Winners

The Wildman Medal was established in 1979 in partnership with the CPA firm of Deloitte Haskins & Sells. The award is for the publication (article, book, monograph, etc.) published during the preceding three calendar years that has made or is likely to make the most significant contribution to the advancement of the practice of public accounting, including auditing, tax practice, and management consulting services. Winners receive an inscribed medal and $2,500, now provided by the Deloitte

& Touche Foundation. The Wildman Award commemorates John R. Wildman, a partner in the firm of Haskins & Sells from 1918 to 1938, who was also a professor at New York University. Wildman was the first president of The American Association of University Instructors in Accounting, the predecessor name of the group that became the American Accounting Association in 1936. Winners of the award and the works for which they were honored are listed below.

Year	Name	Title	Publisher
1979	Loyd C. Heath	"Financial Reporting and the Evaluation of Solvency"	Accounting Standards Division, AICPA, 1978
1980	Donald A. Leslie Albert D. Teitlebaum Rodney J. Anderson	"Dollar-Unit Sampling, A Practical Guide for Auditors"	Fearon-Pitnam Publishers, Inc. 1979
1981	Wanda A. Wallace	"The Economic Role of the Audit in Free and Regulated Markets (A Teaching Tool)"	Touche Ross & Co. 1980
1982	Theodore J. Mock Jerry L. Turner	"Internal Accounting Control Evaluation and Auditor Judgment"	AICPA
1983	William R. Kinney, Jr.	"Regression Analysis in Auditing"	Journal of Accounting Research, 1982
		"Mitigating the Consequences of Anchoring in Auditor Judgements"	Accounting Review January, 1982
		"The Decision Theory Approach to Audit Sampling"	Journal of Accounting Research Spring, 1979
		"Perception of the Internal and External Auditor as a Deterrent to Corporation Irregularities"	Accounting Review July, 1981
1984	Henry R. Jaenicke	"Survey of Present Practices in Recognizing Revenues, Expenses, Gains, and Losses"	FASB, 1981
1985	William H. Beaver	"Incremental Information Content of Statement 33 Disclosures"	FASB, 1983

1986	Frederick D. S. Choi Gerhard G. Mueller	"International Accounting"	Prentice-Hall, Inc., 1984
1987	Barry E. Cushing James K. Loebbecke	"Comparison of Audit Methodologies of Large Accounting Firms"	AAA 1986 (Research Study #26)
1988	H. Thomas Johnson Robert S. Kaplan	"Relevance Lost: The Rise and Fall of Management Account- ing"	Harvard Business School Press, 1987
1989	Treadway Commission	"Report of the National Commission on Fraudulent Financial Reporting"	1987
1990	A. Rashad Abdel-khalik Ira Solomon	"Research Opportunities in Auditing: The Second Decade"	AAA 1989

Seminal Contribution to Accounting Literature Awards

When Steve Zeff became AAA president in 1985, he suggested the establishment of a new award called the Seminal Contribution Award which was to recognize works that have stood the test of time and have contributed in a fundamental way to later research. The award will be given no more frequently than once every three years, and only one work will be honored in any year. Eligible works for this award must have been published at least 15 years prior to the year in which the award is bestowed. So far, the award has been given twice—first in 1986 and again in 1989. The winners were:

YEAR	NAME	TITLE	PUBLISHER
1986	Philip Brown Ray Ball	"An Empirical Evaluation of Accounting Income Numbers"	Journal of Accounting Research, 1968
1989	William H. Beaver	"Information Content of Annual Earnings Announcements"	Journal of Accounting Research, 1968

Manuscript Contest

The AAA encourages research among its younger members by sponsoring an annual manuscript award which is open to anyone who has earned their doctorate in the past five years, or who have not reached their 31st birthday, whichever is later. Coauthored papers are not eligible. As many as three manuscripts may be selected each year. Winners are listed on the next page.

AAA MANUSCRIPT CONTEST WINNERS 1966 TO PRESENT

Year	Name	Title	Published in *The Accounting Review*
1966	George Benston	"Multiple Regression, Analysis of Cost Behavior"	October 1966
1967	Daniel McDonald	"Feasibility Criteria for Accounting Measures"	October 1967
	Abdellatif Khemakhen	"A Simulation of Management-Decision Behavior: Fund and Income"	July 1968
	William H. Beaver	"Alternative Accounting Measures and Predictors of Failure"	January 1968
1968	Gerald Feltham	"The Value of Information"	October 1968
	Charles Tritschler	"Statistical Criteria for Asset Valuation by Specific Price Index"	January 1969
1969	Russell Barefield	"A Model of Forecast Biasing Behavior"	July 1970
	James Winjum	"Accounting in its Age of Stagnation"	October 1970
1970	James C. McKeown	"An Empirical Test of a Model Proposed by Chambers"	January 1971
1971	A. Rashad Abdel-khalik	"User Preference Ordering Value: A Model"	July 1971
	Theodore J. Mock	"Concepts of Information Value and Accounting"	October 1971
1972	Thomas Hofstedt	"Some Behavioral Parameters of Financial Analysis"	October 1972
	John Dickhaut	"Alternative Information Structures and Probability Revisions"	January 1973
1973	A. Rashad Abdel-khalik	"The Entropy Law, Accounting Data & Relevance to Decision Making"	April 1974
	Richard F. Kochanek	"Segmented Financial Disclosure By Diversified Firms and Security Prices"	April 1974
	Theodore J. Mock	"The Value of Budget Information"	July 1973
1974	Yoshihide Toba	"A General Theory of Evidence as the Conceptual Foundations in Auditing Theory"	January 1975
	Rick Elam	"The Effect of Lease Data on the Predictive Ability of Financial Ratios"	January 1975

Year	Name	Title	Published in *The Accounting Review*
1975	Shyam Sunder	"Properties of Accounting Numbers Under Full Costing and Successful Efforts Costing in the Petroleum Industry"	January 1976
	George Foster	"Accounting Earnings and Stock Prices of Insurance Companies"	October 1975
	E. Daniel Smith	"The Effect of the Separation of Ownership From Control on Accounting Policy Decisions"	October 1976
1976	David C. Hayes	"The Contingency Theory of Managerial Accounting"	January 1977
	George Foster	"Quarterly Accounting Data: Time-Series Properties and Predictive-Ability Results"	January 1977
	Larry L. Lookabill	"Some Additional Evidence on the Time Series Properties of Accounting Earnings"	October 1976
1977	Sanjoy Basu	"The Effect of Earnings Yield on Assessments of the Association Between Annual Accounting Income Numbers and Security Prices"	April 1978
1978	Jerold Zimmerman	"The Costs and Benefits of Cost Allocation"	October 1979
1979	Robert M. Bowen	"Valuation of Earnings Components in the Electric Utility Industry"	January 1981
1980	Michael W. Maher	"The Impact of Regulation of Controls: Firms' Response to the Foreign Corrupt Practices Act"	October 1981
1981	James F. Sepe	"The Impact of the FASB's 1974 Proposal for General Price-Level Adjusted Financial Information on the Security Price Structure"	July 1982
1982	Chee W. Chow	"The Impacts of Accounting Regulation on Bondholder and Shareholder Wealth: The Case of the Securities Act"	July 1983
1983	Chee E. Chow	"The Effects of Job Standard Tightness and Compensation Scheme on Performance: An Exploration of Linkages"	October 1983
1984	Mark Zmijewski	"A Test of the Information Content of Financial Statements Beyond that Contained in Earnings Numbers"	January 1987

Year	Name	Title	Published in *The Accounting Review*
1985	Keith A. Shriver	"An Empirical Examination of the Measurement Error Inherent in the Producer Price Indexes and the Implications for the Financial Reporting of Changing Prices"	January 1987
1986	G. Peter Wilson	"The Incremental Information Content of the Accrual and Funds Components of Earnings After Controlling for Earnings"	April 1987
1987	Zoe-Vonna Palmrose	"A Critical Analysis of Auditor Litigation and Audit Service Quality"	January 1988
1987	Terry Shevlin	"Taxes and Off-Balance Sheet Financing: Research and Development Limited Partnerships"	
1988	John R. M. Hand	"Why Did Firms Undertake Debt-Equity Swaps?"	October 1989
1989	John R. M. Hand	"A Test of the Extended Functional Fixation Hypothesis"	October 1990
1989	Siva Swaminathan	"The Impact of SEC Mandated Segment Data on Price Variability and Divergence of Beliefs"	
1990	Terry Shevlin	"The Valuation of R&D Firms With R&D Limited Partnerships"	January 1991

Summary and Conclusion

As was true in many other areas, the beginning of this quarter century, particularly the year 1966, marked a major turning point in AAA research activities. For one thing, *The Accounting Review* became a peer reviewed journal. In addition, a new outlet for research, a monograph series, became available. The year 1966 also marked the beginning of the end for committee research, and individual research came to the forefront. The year 1966 also marked the first year that the AAA/AICPA Outstanding Contribution to Accounting Literature Award was given. Thus, a new era had begun; research grew in importance, and it was a different type of research. At about the same time the AACSB mandated a doctoral requirement for accounting professors. This too resulted in increased accounting research, and an increased need for research outlets. The AAA responded to this need for increased outlets by publishing many monographs and starting new journals. When the journals published by the sections are considered, the AAA now publishes eight different journals, in contrast to the one that was published in 1966. The journals published by the sections are discussed in Chapter 5. By the end of the third quarter century, these research outlets have become the primary research contribution of the AAA. The AAA responded to the increased number of publications with a new desktop publishing system in 1989. Pat Calomeris of the AAA staff is responsible for the in-house typesetting of most national and section publications, including journals, newsletters, and other publications. The availability of this system has enabled the Association to maintain its publication pace, and keep costs at a minimum.

Whereas the third quarter century began with the Association providing research grants to induce scholars to author monographs, that leadership in research funding was eventually lost to the CPA firms whose foundations now fund hundreds of thousands of dollars worth of academic research each year. The availability of funds from organizations such as the KPMG Peat Marwick Foundation and the Arthur Young Foundation have resulted in increased levels of research, particularly in the areas of auditing and taxation. This increased research has led to a need for even more publication outlets. In summary, the AAA has not only been responsible for an increase in the quantity of accounting research, it has also responded to the need for more research outlets that was dictated by external forces.

The individuals at the heart of the AAA's research efforts have been the directors of research and the editors of the Association journals. These individuals have not been alone, however, as research has also been encouraged in other ways. For instance, the annual meetings (discussed in Chapter 3) have also been research motivators since the early 1970s, as have the regional meetings. In addition, most of the sections have research outlets in the form of either journals or monographs. In many respects, the AAA has gotten a great deal of "bang for its buck" in the area of research. The monographs are not particularly costly, *The Accounting Review* operates on the funds received from sales of advertising, and the annual meeting is a profit center. Thus, accounting research has been encouraged at little cost to the membership. Many of the research directors have expressed doubts about their accomplishments, but in reality the accomplishments have always been there, albeit unmeasurable.

References

Abdel-khalik, A. Rashad, "An Analysis of the Attitudes of a Sample of the AAA Members Toward *The Accounting Review*," *The Accounting Review*, July, 1976, pp. 604-616.

Caplan, Edwin H., and Charles H. Griffin, "*The Accounting Review*—A Readership Survey," Mimeo (The American Accounting Association, 1972).

Chatfield, Michael, "*The Accounting Review's* First Fifty Years," *The Accounting Review*, January, 1975, pp. 1-6.

Dyckman, Thomas R., Interview by Terry K. Sheldahl, December 9, 1989.

Hakansson, Nils H., "Where We Are in Accounting: A Review of 'Statement on Accounting Theory and Theory Acceptance,'" *The Accounting Review,* July, 1978, pp. 717-725.

Kinney, William R., Letter to Dale L. Flesher, March 25, 1991.

Minutes of the American Accounting Association Executive Committee Meetings, 1965-1990.

Mock, Theodore, Letter to Dale L. Flesher, March 27, 1991.

Previts, Gary John and Bruce Committe, *An Index to the Accounting Review, 1926-1978* (Sarasota: American Accounting Association, 1980).

"Report of the Committee to Examine the Editorial Policy of *The Accounting Review* and to Consider Additional Publication Outlets," Mimeo (American Accounting Association, July 30, 1976).

Skousen, K. Fred, Letter to Dale L. Flesher, March 28, 1991.

Statement on Accounting Theory and Theory Acceptance (Sarasota: American Accounting Association, 1977).

Sterling, Robert R., Letter to Dale L. Flesher, March 25, 1991.

Sunder, Shyam, Letter to Dale L. Flesher, March 27, 1991.

Zeff, Stephen A., Interview by Dale L. Flesher, December, 1989.

Zeff, Stephen A., "1977-78 Report of the Editor," *Accounting Education News,* November, 1978, p. 7.

Zeff, Stephen A., Letter to Dale L. Flesher, April 15, 1991.

Zlatkovich, Charles, Interview by Dale L. Flesher, March, 1990.

CHAPTER 9
SUMMARY

Summarizing the history of the AAA during the past quarter century in a few paragraphs is a near impossible task. Perhaps the easiest conclusion is to say that the past quarter century has been one of pluralism as individuals, committees, sections, and regions have all grown and developed. Both the breadth and depth of activities at the national, sectional, and regional levels are impressive. The growth in sectional and regional activities has resulted in an array of high quality programs taking place under the total umbrella of the AAA. Although all of the regions existed in 1966 except for the Mid-Atlantic Region, their annual meetings were quite small. In most cases, only three or four speakers were involved in the programs. Today, regional meetings attract hundreds of members, and the number of program participants is often close to a hundred or more. Similarly, the advent of sections in 1976 has resulted in increased opportunity for membership participation. There are now eleven special interest sections, and the activities of most would probably exceed the activities of the entire Association in 1966.

In 1966, the Association published one journal and no newsletters; today, it publishes eight journals and 12 newsletters. Specialized conferences were nonexistent in 1966. By 1991 there are ten different Association-wide conferences held on either an annual or biennial basis, plus others sponsored by the sections. Such tremendous growth in a short period of time has created some stresses within the organization, but these have been relatively minor given the extent of the Association's activities. Much of the problems that could have arisen have been alleviated because of the communication brought about by the advent of the Council in 1978. Regardless of the minor conflicts that have occurred over the years, one cannot help but be impressed by the obvious vitality of the AAA as a total organization.

The activities surrounding the annual meeting are indicative of the growth of the organization. The annual meeting has long been the highlight of member activities, but in 1966 the meeting was only two days in length. In 1990 and 1991, the three-day meeting has been preceded by another day of continuing education activities, and still another day of committee meetings. Many members now arrive on Saturday and stay until the following Thursday morning. Even with the extended time period, many members find it difficult to schedule all of the necessary activities brought about by involvement in multiple sections and special activities.

As activities have increased, so have the costs of these activities. At the same time, total membership has been declining or staying constant due to the loss of practitioner members. Consequently, budget deficits have been the norm in recent years. However, the dues increase which took effect in 1991 should relieve the budget problem at least temporarily. Despite high inflation, dues increases have been few and far between during the past quarter century due to the increasing donations from CPA firms and industries. Only twice prior to 1991 have dues been increased from the $10 charged in 1966. The 1974 change was the highest in percentage terms going up to $25. In 1985, the increase was to $45. The recent increase brings the 1991 level to $65. In 1965, contributions to the Association amounted to about $1,000. In 1990, the figure is close to $1 million. Such support from practitioners indicates that the AAA must be doing something right. Given this overwhelming increase in monetary support, the declining number of practitioner members can be somewhat overlooked. It could be that many practitioners do not want to read our journals, but they like the work we are doing.

When Al Arens took office as the 75th president of the Association in August 1990, he outlined what he felt were the fourteen most important decisions that the Association's leaders had made during the past quarter century. These were divided into two groups—ten that have already proved to be important, and four that are likely to become significant in upcoming years. Those decisions which have already proved to be significant were:

1. Hiring Paul Gerhardt as executive secretary.
2. Fostering sections.
3. Starting the doctoral consortiums.
4. Cosponsoring the Trueblood Seminars.
5. Establishing the Outstanding Educator Awards.
6. Establishing the publication *Accounting Education News.*
7. Establishing Council and changing the nominations process.
8. Sponsoring continuing education activities.
9. Establishing the publications *Issues in Accounting Education* and *Accounting Horizons.*
10. Purchasing our own building.

In addition, Arens listed the following four decisions as likely to make major contributions in coming years:

1. Forming the Accounting Education Change Commission.
2. Establishing four new conferences in recent years:
    a. Professionalism and Ethics Seminar
    b. Financial Reporting Issues Seminar
    c. Senior Faculty Consortium
    d. Corporate Accounting Policy Seminar
3. Holding an Executive Committee meeting in the Netherlands in late 1989, indicating the fact that AAA is indeed an international organization.

4. Allowing the Executive Committee to speak officially for the Association [Arens, 1990].

It would be difficult to dispute the merits of any of the above decisions; these have been the major activities of the Association in the past 25 years. Arens intentionally listed the hiring of Paul Gerhardt as number one on his list, because that is probably the most significant event in the history of the organization. All of the past presidents interviewed for this volume have eulogized Paul Gerhardt in the most flattering terms. When this has been done at an annual meeting, some might interpret such praise as nothing more than the appropriate remarks to be made for the occasion, but how wrong they are. It is unfortunate that only a relative few have the privilege of working closely with him, for those who do soon begin to appreciate the outstanding job that he is doing for our organization. The members of AAA are fortunate to have Paul Gerhardt as executive secretary. In addition to Paul, the Association has a retinue of eleven other talented people in the national headquarters. In particular, Marie Hamilton should be mentioned for her nearly 20 years of service to the organization. Although most of the items on the above lists are of such importance that their significance cannot be questioned, some might question the importance of holding the Executive Committee meeting in the Netherlands, since the Association was already an international organization (21% of 1990 academic members resided outside of the United States). Perhaps holding the meeting overseas was merely acknowledgement by the Executive Committee of what many members, particularly those in the International Accounting Section, have known for years.

As Alvin Arens also pointed out in his inaugural speech, the Association never stops doing anything (well, hardly ever). Thus, all of the programs begun during the past quarter century are likely to continue. Therefore, the next quarter century would appear to be an active one for the Association as new leaders will start new programs, and committees and sections will continue those of the past. The AAA's fourth quarter century will begin with a practitioner president—something that most members interviewed said would never happen again. Even when practitioners made up 75 percent of the membership, it was accepted that the president would normally be an academic. Now that practitioners make up only a small percentage of the membership, one of their numbers (albeit one with a Ph.D.) is elected to the highest office. Thus, the future is difficult to predict. The American Accounting Association has matured, but is still growing and changing. Given this stability combined with growth, the volume commemorating the 100th anniversary will likely be far lengthier than this one.

Reference

Arens, Alvin A., "Presidential Address: Celebration, Evaluation and Rededication," *Accounting Horizons*, December, 1990, pp. 88-96.

1990-91 Executive Committee: (front row, l to r) Corine T. Norgaard, Arthur Wyatt, Alvin A. Arens, David Wilson; (back row, l to r) Daniel Collins, Nicholas Dopuch, Joseph Schulta, Jr., Wanda A. Wallace, John K. Simmons, and Mark Wolfson.